Findings from Production Management Research

Series Editor

Peter Burggräf, Intl Production Engg & Management, Universität Siegen, Kreuztal, Nordrhein-Westfalen, Germany

In the dynamic realm of industrial operations, production management remains the core for organizational efficiency. This series is a collection of doctoral thesis delving into the latest insights and innovations shaping production management, offering a comprehensive exploration of emerging trends and best practices. From the integration of advanced digital technologies to the effects of sustainability initiatives, this series covers key facets driving the evolution of modern production management. Drawing upon cutting-edge research and practical case studies, this collection is an indispensable resource for industry professionals, academics and students seeking to navigate and capitalize on the dynamic landscape of modern production management.

Benjamin Heinbach

Reinforcement Learning-Based Planning of Factory Layouts

Benjamin Heinbach
Universität Siegen
Siegen, Germany

ISSN 3005-1649 ISSN 3005-1657 (electronic)
Findings from Production Management Research
ISBN 978-3-658-51553-9 ISBN 978-3-658-51554-6 (eBook)
https://doi.org/10.1007/978-3-658-51554-6

This Springer Vieweg imprint is published by the registered company Springer Fachmedien Wiesbaden GmbH, part of Springer Nature.
The registered company address is: Abraham-Lincoln-Str. 46, 65189 Wiesbaden, Germany

Supervisors Foreword

In the dynamic realm of industrial operations, production management remains the core for organizational efficiency. This series is a collection of doctoral thesis delving into the latest insights and innovations shaping production management, offering a comprehensive exploration of emerging trends and best practices. From the integration of advanced digital technologies to the effects of sustainability initiatives, this series covers key facets driving the evolution of modern production management. Drawing upon cutting-edge research and practical case studies, this collection is an indispensable resource for industry professionals, academics and students seeking to navigate and capitalize on the dynamic landscape of modern production management.

Danksagung

Eine 3120 Tage lange Reise findet nun endlich ein würdiges Ende.

Viele prägende Gesichter haben mich auf diesem Weg begleitet, und ich möchte die Gelegenheit nutzen, einigen von ihnen meinen Dank auszusprechen.

Diese Dissertation zur Planung von Fabriklayouts mithilfe von Reinforcement Learning entstand teilweise während meiner Tätigkeit als wissenschaftlicher Mitarbeiter am Lehrstuhl für International Production Engineering and Management (IPEM) an der Universität Siegen, und teilweise während meiner Zeit als Systementwickler IoT bei der Achenbach Buschhütten GmbH & Co. KG.

Zuallererst möchte ich Prof. Dr.-Ing. Peter Burggräf, Inhaber des IPEM-Lehrstuhls, für die Möglichkeit zur Promotion sowie für seine fortwährende Unterstützung und Ermutigung danken – die auch nach meinem Weggang vom Lehrstuhl anhielt – und für sein Vertrauen in mich als seinen allerersten wissenschaftlichen Mitarbeiter in Siegen, trotz meines durchaus etwas exotischen Hintergrunds.

Ich danke außerdem Prof. Dr.-Ing. habil. Matthias Schmidt für seine Bereitschaft, als Zweitgutachter zu fungieren. Prof. Dr. Ralph Dreher danke ich dafür, dass er den Vorsitz der Promotionskommission kompetent übernommen hat, für sein Feedback zu meiner Arbeit sowie für die unkomplizierte Durchführung des Prüfungsverfahrens. Und Prof. Dr.-Ing. Axel von Hehl danke ich für seine Tätigkeit als Prüfer in der Promotionskommission.

Viele meiner ehemaligen Kolleginnen und Kollegen am IPEM hatten einen maßgeblichen Einfluss auf diese Arbeit oder die Prüfung, und ich bin ihnen von Herzen dankbar. Daher möchte ich besonders die Zusammenarbeit mit Herrn Till Saßmannshausen, Dr.-Ing. Johannes Wagner, Dr.-Ing. René Sauer, Dr.-Ing.

Fabian Steinberg, Philipp Nettesheim, Maximilian Schütz und Norman Müller hervorheben. Mein besonderer Dank gilt allen studentischen Hilfskräften sowie den Studierenden, die mit ihren Abschlussarbeiten und ihrer Unterstützung zu dieser Arbeit beigetragen haben.

Mein tiefster und herzlichster Dank gilt meinen Eltern Antje und Thomas Koke für ihre bedingungslose Unterstützung auf meinem Lebensweg.

Und schließlich möchte ich meinen wunderbaren Kindern, Oscar Jaro und Fiona Linnea, danken. Ihr habt nicht nur die Farben in dieses Buch gebracht, sondern bringt jeden Tag Farbe und Licht in mein Leben. Ich liebe euch beide sehr.

Wenden Benjamin Heinbach
February 2026

Abstract

Facility layout planning is a core discipline in production management, directly shaping operational efficiency, material flow, and cost structures. Despite its criticality, facility layout planning presents a complex combinatorial problem, often approached through heuristics or metaheuristics that lack scalability and adaptability. This dissertation investigates the use of (Deep) Reinforcement Learning (DRL) to automate and enhance layout planning by conceptualising facility layout planning as a Markov Decision Process (MDP). This thesis demonstrates that DRL agents – trained solely through interaction feedback without domain-specific input – can autonomously generate layout configurations that significantly reduce material handling costs and generalise across varying problem instances, thus demonstrating DRL's viability as a scalable and adaptive resolution technique for facility layout planning. Building on the conceptual parallel between human iterative layout adjustment and Reinforcement Learning processes, this research follows a Design Science Research paradigm of experimental artefact design. It unfolds over four peer-reviewed publications. Initially, through a systematic literature review, the work identifies a significant underrepresentation of RL in layout optimisation research (1). An open-source simulation environment was developed to operationalise this approach, allowing flexible training and benchmarking of RL agents across various facility layout planning instances (2). Experimental results demonstrate that DRL agents, trained solely through visual interaction feedback without domain-specific features, can achieve substantial improvements in material handling costs (3). Further studies on model generalisation reveal promising transferability across small layouts, while larger instances present new challenges (4). Beyond the experimental contributions, this work opens a path toward AI-driven factory planning tools that can potentially reduce planning effort, improve layout quality, and ultimately enable more responsive and data-driven production system design in dynamic industrial environments.

Zusammenfassung

Die Fabriklayoutplanung (Facility Layout Planning) ist eine Kerndisziplin des Produktionsmanagements, die maßgeblich die operative Effizienz, den Materialfluss und die Kostenstrukturen beeinflusst. Fabriklayoutplanung stellt ein komplexes kombinatorisches Optimierungsproblem dar, das häufig durch Heuristiken oder Metaheuristiken adressiert wird, denen es jedoch an Skalierbarkeit und Anpassungsfähigkeit mangelt. Diese Dissertation untersucht den Einsatz von Deep Reinforcement Learning (DRL) zur Automatisierung und Verbesserung der Layoutplanung, indem die Fabriklayoutplanung als Markov-Entscheidungsprozess (MDP) konzeptualisiert wird. Es wird einerseits demonstriert, dass DRL-basierte Agenten – ausschließlich durch Interaktionsfeedback ohne domänenspezifisches Vorwissen trainiert – in der Lage sind, selbstständig Fabriklayouts zu erzeugen, die die Materialtransportkosten signifikant reduzieren, und andererseits auf unterschiedliche Problemgrößen generalisieren können. Anhand dieser Ergebnisse wird das Potenzial von DRL als skalierbare und adaptive Lösungsmethodik für praktische Fabriklayoutprobleme aufgezeigt.

Motiviert von der konzeptionellen Parallele zwischen zumeist manueller, iterativer und zumeist menschlicher Layoutplanungsprozesse und aktuellen Lernerfolgen des Reinforcement Learnings werden in dieser Arbeit auf Basis eines der Design Science Research Methodik über vier peer-reviewte Publikationen hinweg experimentell folgende Artefakte zur Problemlösung entwickelt: Eine systematische Literaturanalyse identifiziert zunächst eine deutliche Unterrepräsentation von RL in der Fabriklayoutplanungsforschung (1). Die darin herausgearbeiteten Konzepte der Fabriklayoutplanung wurden in MDPs überführt und in einer Open-Source-Simulationsumgebung operationalisiert, um ein flexibles Training sowie Benchmarking von RL-Agenten über verschiedene Fabriklayoutproblem-Instanzen hinweg zu ermöglichen (2). Mithilfe der Umgebung durchgeführte experimentelle Ergebnisse belegen, dass DRL-Agenten, die ausschließlich durch visuelles Interaktionsfeedback ohne domänenspezifische Merkmale trainiert wurden, erhebliche Verbesserungen der Materialflusskosten erzielen können (3). Darauf aufbauende Studien zur Modellgeneralisierbarkeit zeigen eine vielversprechende Übertragbarkeit auf kleinere Layoutprobleme, während größere Instanzen neue Herausforderungen mit sich bringen, die der weiteren Anpassung des zugrunde liegenden MDP bedürfen (4).

Über die experimentellen Beiträge hinaus ebnet diese Arbeit den Weg für KI-gestützte Fabrikplanungswerkzeuge, die künftig potenziell Planungsaufwände reduzieren, die Layoutqualität verbessern und letztlich eine reaktionsfähigere und datengetriebene Gestaltung von Produktionssystemen in dynamischen Industrieumgebungen ermöglichen können.

Contents

Symbols and Abbreviations

Abbreviation	Description
A2C	Advantage Actor Critic
A3C	Asynchronous Advantage Actor-Critic
ACO	Ant Colony Optimisation
AGV	Automated Guided Vehicle
AHP	Analytic Hierarchy Process
AI	Artificial Intelligence
ALDEP	Automated Layout Design Program
ANN	Artificial Neural Network
ANP	Analytic Network Process
CORELAP	Computerised Relationship Layout Planning
CRAFT	Computerised Relative Allocation of Facilities Technique
DDPG	Deep Deterministic Policy Gradient
DFLP	Dynamic Facility Layout Problems
DQN	Deep Q-Networks
DRL	Deep Reinforcement Learning
DSR	Design Science Research
FBS	Flexible Bay Structure
FLP	Facility Layout Problem
GA	Genetic Algorithms
GNN	Graph Neural Network
HVAC	Heating-Ventilation-Air-Conditioning
IS	Information Systems

JSSP	Job-Shop Scheduling Problem
MCTS	Monte Carlo Tree Search
MDP	Markov Decision Process
MHC	Material Handling Cost
MILP	Mixed-Integer Linear Programming
ML	Machine Learning
NP	Non-deterministic polynomial-time
OFLP	Open Field Layout Problem
PPO	Proximal Policy Optimisation
PSO	Particle Swarm Optimisation
QAP	Quadratic Assignment Problem
QSP	Quadratic Set Covering Problems
ReLU	Rectified Linear Unit
RGB	Red-Green-Blue
RL	Reinforcement Learning
SA	Simulated Annealing
SAC	Soft Actor-Critic
SARSA	State-Action-Reward-State-Action
SFLP	Static Facility Layout Problems
SLR	Systematic Literature Review
SOM	Self-Organizing Maps
STS	Slicing Tree Structure
TD	Temporal Difference
TD3	Twin-Delayed DDPG
TOPSIS	Technique for Order Preference by Similarity to Ideal Solution
TRPO	Trust Region Policy Optimisation
TS	Tabu Search
WIP	Work-In-Process

List of Figures

List of Tables

1 Introduction

Facility layout planning is a crucial aspect of production management that involves arranging various functional elements within a facility to optimise efficiency and productivity. It can be defined as a systematic, goal-oriented process structured into successive phases and carried out with the aid of methods and tools, encompassing the planning of a factory from the initial idea to its construction and production ramp-up (VDI, 2011). Factory planning presents a unique challenge, as determining future activities and solutions must be completed well in advance. Planning thus entails designing not only the physical layout but also the underlying production processes long before they become operational (Grundig, 2021). This dilemma highlights the core tension in factory planning: creating production systems that must remain operational for decades while remaining adaptable to shifting market demands (Schuh et al., 2011).

The significance of facility layout planning in production management is underscored by its impact on operational efficiency. A well-designed layout can lead to substantial reductions in operational costs, with studies indicating potential savings of 10% to 50% in total operating costs through effective layout design (Besbes et al., 2020; Madhusudanan Pillai et al., 2011; Sun et al., 2018; Tompkins et al., 2010), or cost increase risks of 36% (Ripon et al., 2013), respectively. Moreover, the layout directly influences productivity metrics, as it affects the flow of materials and personnel within the facility (Kovács and Kot, 2017; Liu and Meller, 2007).

Facility layout planning is further characterised by complex problem sets that often lack exact solutions due to the multifaceted nature of the variables involved. As such, the Facility Layout Problem (FLP) is recognised as NP-hard, indicating

B. Heinbach, *Reinforcement Learning-Based Planning of Factory Layouts*, Findings from Production Management Research ,
https://doi.org/10.1007/978-3-658-51554-6_1

that finding optimal solutions is computationally challenging and often requires heuristic or approximating methods (Salas-Morera et al., 2021). This complexity arises from the need to consider numerous factors such as spatial constraints, workflow efficiency, safety, and cost implications, which can vary significantly across different production environments (El-Rayes and Khalafallah, 2005; Sun et al., 2018). The iterative nature of layout planning, where adjustments are made based on performance feedback and changing operational needs, further complicates the process, reinforcing the notion that exact solutions are rarely attainable (Zeng et al., 2023).

In summary, facility layout planning is an integral subset of production management that involves navigating significant, complex problem sets without straightforward solutions. The interplay of various factors such as efficiency, safety, and cost necessitates a flexible and adaptive approach to layout design, often relying on heuristic methods to arrive at satisfactory, albeit not always optimal, solutions (Miç, 2022; Peron et al., 2020).

1.1 On the Importance of AI-Assisted Layout Planning

Extensive research on FLPs dates back to the 1950s, when seminal work by Koopmans and Beckmann (1957) formulated layout challenges as Quadratic Assignment Problems. Subsequent studies have aimed to refine these formulations or propose resolution methods—whether exact, heuristic, or meta-heuristic—to advance the boundaries for objective functions such as material handling cost.

Although FLP research typically focuses on quantitative metrics like inter-facility transport intensity (Sharma and Singhal, 2016), successful layout solutions also hinge on qualitative considerations that extend beyond measurable indicators (Amar and Abouabdellah, 2016; Burggräf, Adlon et al., 2021; Hosseini-Nasab et al., 2018). In practice, a dedicated project team balances both quantitative and qualitative factors through multi-criteria decision-making approaches such as the analytic hierarchy process (Saaty, 1977) or TOPSIS (Hwang, 1981). While these methods can guide layout development, their reliance on expert input introduces potential biases (Odu, 2019), leaving them susceptible to personal or organisational agendas.

Besides the subjective nature of manual planning results, the process is time-consuming and resource-intensive. To address this, a multitude of automated layout planning approaches have been developed in the past decades. These include exact methods which tend to fail on larger problems because of an exponentially growing computation time, stochastic methods, heuristic solutions, and meta-heuristics that approximate near-optimal solutions (Drira et al., 2007). However, such algorithmic approaches typically fail to incorporate soft or qualitative criteria that are less tractable and hard to quantify.

Furthermore, a comprehensive overview of FLP research reveals a notable lack of Machine Learning (ML) integration (Bouramtane et al., 2024; Drira et al., 2007; Hosseini-Nasab et al., 2018; Pérez-Gosende et al., 2021; Renzi et al., 2014), a gap that seems incongruous given the growing prominence of ML in production research (Burggräf et al., 2018; Cadavid et al., 2019; Cioffi et al., 2020; Kang et al., 2020; Panzer and Bender, 2022; Weichert et al., 2019). One subclass of ML is Reinforcement Learning (RL), or, when combined with deep neural networks for feature extraction, Deep Reinforcement Learning (DRL). Since the groundbreaking results of DRL in mastering Chess, Go, and Atari games, research interest in the field has surged, extending far beyond game-playing into real-world applications. The success of AlphaGo (Silver et al., 2016) and deep Q-networks (Mnih et al., 2015) demonstrated DRL's ability to solve complex sequential decision problems with high-dimensional state spaces, sparking advancements in robotics, autonomous systems, and industrial optimisation. As a result, DRL is now being actively explored for applications in various areas, where intelligent agents must learn adaptive policies in dynamic and uncertain environments.

Many production management tasks involve sequential decision processes with vast state and action spaces, a setting that is well-suited for Markov decision process (MDP) formulations and, consequently, DRL. Indeed, DRL has delivered promising results across multiple production and operations research domains, notably production scheduling (Burggräf et al., 2020; Gabel and Riedmiller, 2008; Kuhnle et al., 2019), robotics, e.g. bin-picking (Moosmann et al., 2021; Stuke et al., 2024), automated assembly (Beltran-Hernandez et al., 2020), or intricate motor tasks such as solving a Rubik's Cube (OpenAI et al., 2019), to name but a few. Such achievements underscore RL's demonstrated capabilities in solving sequential decision tasks and, by extension, support its application to FLPs, which share the combinatorial complexities.

The idea for pursuing this thesis has been drawn from the seminal paper of Mnih et al. (2015), where the DRL agent learns to play Atari games by observing only the on-screen visual data (i.e., pixels) and learning which moves maximise its eventual score through experimentation with different actions and the observation of the outcomes, ultimately achieving higher results than an expert human player. The mechanism of mastering a video game is remarkably similar to a human planner manually rearranging machines, workstations, and storage areas on a 2D layout as witnessed by the lead researcher in a variety of industrial factory planning projects: the planner(s) visually inspect(s) how each move affects material flow or distances, seeking to balance various constraints and ultimately to achieve better cost or performance metrics. Much like an RL agent must infer both the "rules" of the game and a winning policy through repeated trials, human factory planners test incremental modifications and assess how these changes impact the overall layout quality. In both contexts, feedback loops drive learning. For the Atari-playing RL agent, reward signals originate from in-game scoring or progress. In contrast, for layout planning, signals may include reduced material handling costs, shorter travel distances, or improved worker safety.

This leads to the key driving motivation of this work: "*If DRL-based systems can match human performance in limited domains, such as Go, chess, or Atari games, by merely learning from visual input; DRL-based systems should also be able to match the performance of human planners of facility layouts as well.*" Fig. 1.1 below represents this idea graphically.

By introducing automation through RL, one can expect the formerly manual process of trial and error in factory layout planning to be accelerated, harnessing computational power to explore numerous potential layouts and systematically improve the arrangement with each iterative step. Such a perspective opens the possibility of more robust, data-driven layout decisions, ultimately alleviating the burdens of traditional, expert-based planning methods and paving the way for more flexible, high-performing factory configurations.

However, at the outset of this work, as sections 2 and 3 will outline, the knowledge base to bridge DRL and FLP research was scarce. While there was some evidence of the use of ML in FLP, notably in the area of unsupervised learning with so-called self-organising maps, DRL had been, by and large, unknown to the problem domain of FLP. Although DRL had been successfully demonstrated to be of use in neighbouring fields of production management, such as scheduling, the principles had seemingly not yet been overcarried to FLP. This thesis will address this objective and is structured as follows: After defining the scope and objectives of the thesis and deriving the guiding research questions, section two summarises the basics of Facility Layout Problems and Reinforcement Learning. Sections three to six cover four peer-reviewed publications and thus represent the cumulative part of this thesis. The four publications are reproduced with standardised formatting and citation style. The figures have been kept in their original style. Finally, a critical reflection and summary are given in the last two sections.

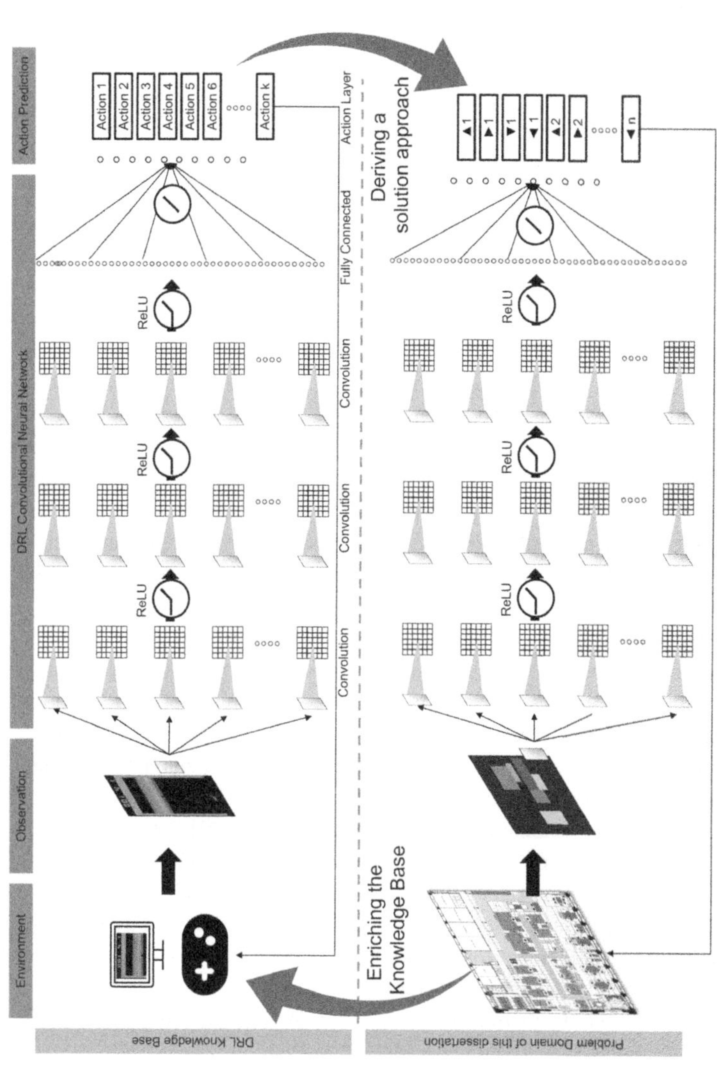

Fig. 1.1 Conceptual overview motivating this dissertation: if DRL agents can learn to play computer games from visual input, they should be able to learn to design layouts as well

1.2 Research Methodology

1.2.1 Methodological Considerations

Before conducting scientific research, it is essential to establish a foundational assumption about epistemology and ontology, as these shape how knowledge is constructed, what qualifies as valid evidence, and how reality is understood, ensuring coherence and methodological rigour in the study.

Epistemology is "the study of the criteria by which we can know what does and does not constitute warranted, or scientific, knowledge" (Johnson and Duberley, 2000, pp. 2–3). There are several positions within the field of epistemology, the two most prominent of those are *positivism* and *interpretivism* (Bryman, 2012). Hart (2004) describes positivism as the philosophical standpoint that accepts only measurable, sensorially perceivable data to create knowledge in an objective, unbiased manner, thereby describing a universal reality consisting of true or false facts. Interpretivism, on the other hand, asserts that social sciences and natural sciences are fundamentally different from each other, and social research requires methods that capture, observe, and explain human behaviour, which in turn shape social reality (Hart, 2004). Enter *realism*, a third epistemological orientation to mediate between positivism and interpretivism. Realist research assumes that objects such as organisations and culture exist and can be subjected to analysis (Gray, 2021). For realists, there exists not only one universal reality, but several layers to it, which can all be revealed systematically and are not exclusively made up of an object-subject dualism, but also by generative mechanisms that link causes and effects more closely (Chia, 2002). Perception of the nature of said reality, however, seems arbitrary: although Bryman (2012) and Chia state that "reality exists and acts independently of our observations" (2002, p. 11), Gough et al. explicitly write that *critical* realism "assumes that our perceptions and beliefs mediate our knowledge of reality" (2012, p. 41).

The second large field of philosophical considerations is ontology, which is "concerned with the nature of social entities" (Bryman, 2012, p. 32) and can be subdivided into *objectivism* (existence of an external reality that is independent of human consciousness), *subjectivism* (social actors give meaning to objects based on dreams, beliefs and assumptions) (Gray, 2021), and *constructivism* (reality is constructed through the interactions of social actors with the world) (Bryman and Bell, 2019).

In essence, the key ontological discussion centres around the question of whether a universal truth exists that can be perceived and thus explained, or if all truths are merely subjective, externalised reflections of values and beliefs embedded in every individual, forged together through social interaction. Epistemology, in turn, is concerned with the study of knowledge—how we know what we know—and the methods used to validate and expand that knowledge.

In engineering and computer science research, epistemology involves questions about the reliability of algorithms, the justification of design choices, and the validity of models or simulations in representing real-world phenomena. Ontology, on the other hand, addresses the nature of reality and the structure of the entities being studied or modelled. In these fields, ontology involves defining the fundamental components of a system (e.g., nodes, edges, data objects) and their relationships, which is especially critical in areas such as artificial intelligence, database design, and systems modelling. While these dilemmas are, by and large, a heated debate among the social, economic, and possibly natural sciences, the research objective at hand warrants adopting a similar perspective in engineering and computer science research as well.

The design science in information systems research (DSR) framework by Hevner et al. (2004), see Fig. 1.2, and the adherent research guidelines, while emphasising people and organisations as environmental inputs to the artefact design process, have a positivist epistemological assumption (a similar assessment can be found in (Marshall and McKay, 2005)). Niehaves (2007) opposes this implicit assumption and argues that DSR is not only a positivist domain but also open to alternative epistemologies. To him, adopting an interpretivist perspective on the DSR guidelines is crucial, particularly regarding the problem domain side, where there may not be a definitive ground truth to the problem description when multiple individuals are involved. Likewise, in the preliminary works for Koke and Moehler (2019), the author argues that a more qualitative and inductive approach should guide research on project management topics, as projects constitute temporary organisational artefacts whose components are less likely to be influenced by external factors rather than the individuals who shape the organisation through their interactions.

However, another interesting consideration emerges if one shifts the perspective from the researcher to the artefact itself and somewhat endows it with a soul. Brey (2005), in search of an epistemological standard for human-computer interaction, argues that the relationship between humans and computers has historically been epistemic in a way that computers extended human cognition, and created hybrid cognitive systems consisting of a human processor and an artificial processor that jointly process information. Only at that time, according to him, has this relationship been supplemented with an ontic function, through simulating virtual and social environments that extend beyond the physical environment and thus become portals to worlds we inhabit. If one now includes the position of

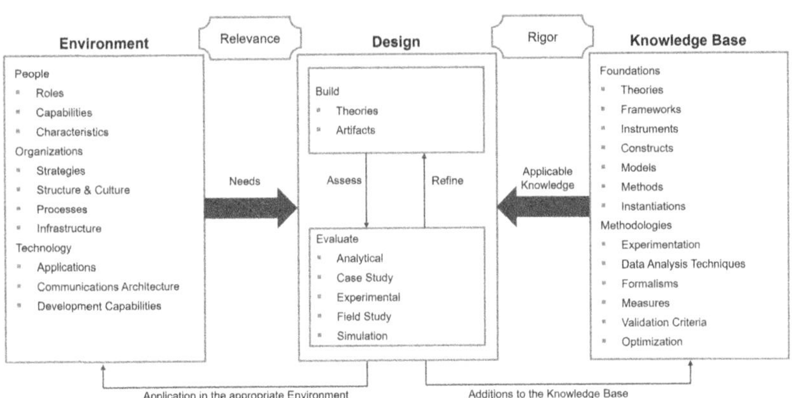

Fig. 1.2 Design Science Research framework (vom Brocke, 2020) adapted from Hevner et al. (2004)

Saßmannshausen and Heinbach (2025) who hypothesised that the scope of cognitive enhancement we already leverage in private and work life should grant us the rank of cyborgs, the conundrum arises whether Brey's assumption of tandem information processing between humans and computers persists.

As the first section has outlined, the search for an AI-based automated planning tool for factory layout has been motivated by the progress of RL in learning sequential decision tasks at a human level (Mnih et al., 2015). These advances may have been greatly influenced by neighbouring fields and their respective success stories, such as deep learning systems outperforming human benchmarks in image recognition (He et al., 2015). These milestones would not have been possible without extensive labelling work on previously unlabeled data for the neural networks to adjust their weights. Likewise, RL systems surpassing humans in a limited domain required the modelling of training environments beforehand. This raises the question at what point humans become the epistemic functions for self-learning computing systems. Moreover, in an extrapolation of this train of thought, how close does a self-learning (in this case RL-) system that learns to extract the general structure of a class of problems approximate the epistemological, interpretivist doctrine of "Verstehen" (Bryman and Bell, 2019), regardless of whether it can be classified as weak or strong Artificial Intelligence (AI).

The challenge now is to integrate these assumptions, considerations, and perspectives into a unified research approach and research objectives.

1.2.2 Research Method Outline

Design Science Research (DSR) seeks to advance knowledge by creating innovative solutions to address practical challenges (vom Brocke, 2020). Hevner's DSR Framework (2004) offers a structured methodology for conceptualising, conducting, and evaluating DSR efforts through three key elements: **Environment**, **Knowledge Base**, and **Design**.

- The **Environment** defines the problem space, encompassing stakeholders, systems, and technologies. It frames the context for identifying challenges and opportunities, providing the foundation for defining research problems rooted in real-world needs.
- The **Knowledge Base** supports problem-solving through existing theories, tools, and methods, providing both foundational insights and methodological rigour. It guides researchers in leveraging prior knowledge and identifying gaps to inform innovation.
- The **Design** element focuses on the iterative process of creating and refining artefacts, which may include constructs, models, methods, or systems. These artefacts are developed to address the identified problems and are systematically evaluated to ensure their relevance, utility, and effectiveness in practice.

This integrated framework ensures that research is both grounded in real-world challenges and contributes to the broader body of knowledge through the development of innovative artefacts.

The previous section has highlighted that adopting an interpretivist epistemological view on some aspects of DSR, as opposed to traditional positivist thinking, can be beneficial. The research herein assumes that both the problem domain, i.e. mainly the organisations and technology elements in the environment, are artificially constructed micro-realities, by and large created through the interactions of the individuals operating therein and therewith. Hence, it appears prudent to adopt an interpretivist perspective on the problem definition, its relevance, and the subsequent discussion.

Regarding the artefact design perspective, however, a positivist approach aligns with the epistemological foundation of objective, measurable, and reproducible knowledge. Unlike interpretivism, which focuses on subjective meaning-making and context-dependent knowledge, positivism ensures that RL systems are rigorously tested through quantitative measures such as reward functions, convergence rates, and generalisation capabilities. This aligns with the principles of DSR, where artefacts are iteratively refined based on objective evaluations rather than human interpretations. Furthermore, as RL algorithms aim to optimise decision-making in complex environments, their design and validation necessitate a structured, experimental methodology that systematically isolates causal relationships between state, action, and reward. Given that the development of RL algorithms relies heavily on computational simulations and empirical validation, a positivist epistemology provides the necessary methodological rigour to establish reliable and generalisable findings.

To summarise, the knowledge base will be constructed interpretivistically. To achieve this, a narrative systematic literature analysis is performed to map its current state as well as the subsequent addition to it. The artefact design, in particular building the simulation environment for the agents, algorithm training and evaluation will be run positivistically.

For the remainder of this thesis, these considerations will unfold as follows: the knowledge base for our use cases is defined in Chaps. 2 and 3. As 2.1 will highlight, the knowledge base on FLP is vast, given that it has grown steadily for almost seven decades. Therefore, building up the knowledge base will be strictly limited to the elements pertinent to this thesis. Then, experimental knowledge is added in Chaps. 4 through 7. The core of DSR revolves around design activities, where innovative solutions are sought to address the research problem. Here, innovative solutions are created by building upon and extending existing design knowledge. This iterative design process involves building and evaluating activities, refining solutions to achieve optimal outcomes. Finally, the relevance of the artefacts and results to the knowledge base and problem environment is discussed in Chap. 8.

1.3 Research Objectives and Contribution of the Four Publications within this Cumulative Dissertation

This thesis is based on the idea of transferring the aforementioned advances in RL to factory planning. The deficiencies in the current state of research regarding the use of RL in factory layout planning can be summarised as follows:

- RL continues to gain increasing recognition and research interest, but its application to real-world problems remains limited.
- While RL has been proven to yield excellent results for highly complex combinatorial problems, it has not yet been established as a viable approach for FLPs.

- There is currently no significant evidence in the literature that examines the generalizability of RL in the context of FLP.

This thesis is therefore driven by the ambition to bridge these gaps and explore the potential of RL as an effective tool for optimising factory layout planning. The research design in this thesis is guided by the primary question: "*To what extent or subject to which limitations, can RL be a technologically viable solution to facility layout planning?*". The attainment of the research goal relies on the content of four publications that were authored throughout this research project. The details of the publications are visually summarised in Fig. 1.3.

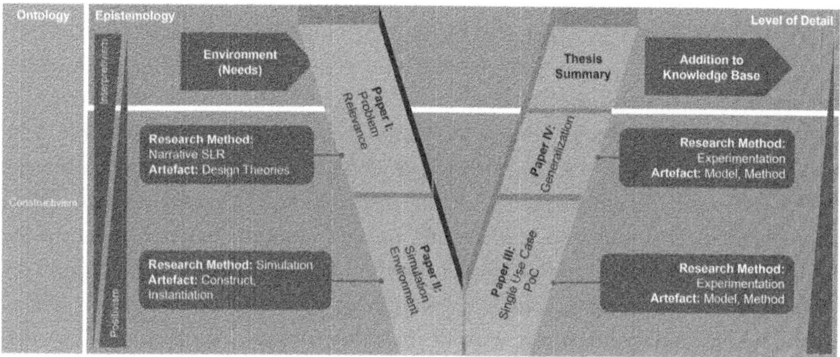

Fig. 1.3 Overview of the four publications and complementary research artefacts within this cumulative dissertation

The first research paper, discussed in Chap. 3, is a comprehensive systematic literature review that examines current methodologies in FLPs and ML for generating and/or optimising facility layouts. Drawing on evidence from neighbouring fields in the production management domain, such as job-shop scheduling, and supported by a previous literature analysis on ML in production management (Burggräf et al., 2018), the review aimed to expand the knowledge base on the application of ML in FLP. Therefore, the research question posed in the paper is "*How have different Machine Learning algorithms been used as resolution techniques for Facility Layout Problems?*". The research methodology of this paper is *analytical*. The systematic analysis reveals that ML is a growing research area. Using a sample of 57 research papers from an originally scoped population of 11,985 references, the Systematic Literature Review (SLR) methodology of Boland et al. (2017) was employed to systematically assess the relevance of

ML for FLP in current research. The results showed that Supervised Learning had been virtually absent, and some works had been published on Unsupervised Learning. Additionally, two contributions in the sample were identified that utilised RL; however, none of these met the inclusion criteria defined in the review. The findings indicated that RL had not yet been adopted as a resolution approach in FLP, thus highlighting the academic need to advance the state of research in the field.

The second publication (Chap. 4) lays the foundation for artefact design, particularly regarding software. In the work, the conceptual framework of the FLP knowledge base, as mapped by the previous SLR, is formalised into a Markov Decision Process (MDP), which serves as the mathematical foundation for solving sequential decision-making problems using RL. The Python-based software library presented in the paper, "gym-flp", encompasses a variety of elements from FLP research, i.e. different discrete and continuous modelling approaches and distance computations. To allow easy access to training RL algorithms on FLP problems, the library leverages OpenAI Gym, making it compatible with many off-the-shelf algorithms. The paper defines various alternatives for the MDP's state space (RGB image vs. vector; discrete vs. continuous) and the action space (discrete, multi-discrete, continuous, or multi-continuous), and performs the requirements engineering of the reward structure based on the FLP literature. Furthermore, in line with the positivistic research agenda outlined earlier, the library implements the flow and, in the case of discrete problems, the distance information of a large number of discrete and continuous problems commonly used in FLP research, providing a benchmark-ready testbed for research on RL in FLPs. The research methodology of paper two is of type *experimental*, by and large driven by unit tests and module tests at code level in the simulation environment.

The third publication (Chap. 5) leverages the software library from the second paper to furnish proof of the validity of the design assumptions made therein and the real-world applicability of the approach. To achieve this, a real-world production scenario within the Smart Demonstration Factory Siegen, which features a workshop comprising twelve functional units, is integrated into the simulation environment. Again, the research methodology is *experimental*, relying on manual algorithm selection, hyperparameter tuning, and analysis of evaluation metrics. From the configuration options in gym-flp, the study utilises a continuous plant representation, where machines have x, y-coordinates, widths, and lengths, and are freely movable within a rectangular plant area. The current layout is the state of the environment and is passed to the RL agent as an RGB image. Gym-flp generally fixes the image size to allow for more straightforward generalisation across different problems. The action space in the study is chosen as (single-)discrete, i.e. one machine is moved at a time with one positional increment (± 1x / ± 1y). To avoid reward gaming of the RL agent, the material handling cost (MHC) of the new layout after taking an action is evaluated against the lowest known value in the current episode. Additionally, the rewards include a component that penalises positioning on restricted areas such as pathways. Finally, an episode is terminated if either structural constraints are violated (i.e., the agent moves a machine outside the plant or machines collide) or if the best-known MHC has not improved for a predefined number of steps. This implementation assumes that the agent will encounter a local optimum. The experiment in the paper utilises three RL algorithms: Deep Q-Networks (DQN), Advantage Actor-Critic (A2C), and Proximal Policy Optimisation (PPO) from the Stable Baselines 3 library. Each algorithm is trained for five million training steps. The results indicate that each algorithm can learn to reduce the layouts' MHC, where PPO showed the highest reduction, while being the only algorithm to produce a valid layout without machines positioned in restricted areas. In essence, this publication demonstrates that RL can improve a layout solely from visual input by learning a policy of sequential machine displacements subject to spatial constraints.

Building on the successful proof of concept presented in the previous paper, the final publication aims to investigate the generalisability of the proposed approach. Analogous to Supervised or Deep Learning solutions, where a model's deployability is tested against new data, RL models should be tested against unseen problems as well. However, the literature on RL in FLP research by and large lacks such a methodology. The experiences from the previous two papers, and in particular the preliminary works on them, have shown that varying the state and action space size drastically inhibits generalisation. While gym-flp effectively handles the state space size issue by clamping the image size to a fixed number, the implementation was still sensitive to different-sized action spaces concerning the number of machines in a problem. To tackle this issue, in a fourth publication (chapter 6) another artefact was added to gym-flp: the new version assumes an upper bound of machines in a plant (62 in this case as this is the largest problem instance implemented) and applies an action mask that disables all invalid actions for more minor problems. To demonstrate the effectiveness of the approach, a comprehensive experiment was conducted: a PPO agent was tasked with learning to sequentially optimise one out of seven problems in the gym-flp environment, by training on random flow matrices and explicitly not the one implemented in the library. The resulting model was stored and re-trained on the following problem instance, and so forth. Ultimately, the final model, incorporating training experience from seven problems, was presented with 100 new random flow matrices, which had not been used in training, for each of the seven problem sizes. The results show that one model can produce a substantial number of feasible layouts with reduced MHC with the given training budget on small layouts, but will underfit on larger problem instances. Nonetheless, generalisation across different problems was successful.

Table 1.1 below presents the target journals for the publications in this thesis, including their Impact Factor (if reported on the journal homepage) and the 2024 Scimago Journal Rank (SJR) according to www.scimagojr.com, as well as the peer-review process for the papers. The metric of the number of revisions does not include previous revisions in other journals that had resulted in a rejection (applicable to papers I and III).

To conclude, this research offers insights into the evolving role of ML, particularly RL, in addressing FLPs through a structured and incremental research agenda. Initially, a systematic literature review highlighted the underrepresentation of RL in FLP applications, identifying a gap in leveraging advanced ML techniques for layout optimisation. Building on this, a software framework was formalised via an MDP and implemented in the "gym-flp" library, providing an accessible, open-source environment for RL experimentation in FLP contexts. The subsequent experimental studies validated the conceptual design by applying RL to a real-world case. Finally, the work explored generalisability, proposing architectural modifications to accommodate varied problem sizes. The cumulative results of this thesis emphasise the potential of RL algorithms in optimising factory layouts in the factory planning process, laying the groundwork for future research in this area.

Hevner et al. (2004) proposed seven guidelines to ensure rigour and relevance in DSR within the field of Information Systems, providing a framework for evaluating and conducting DSR projects. Table 1.2 below summarises how the publication results relate to these guidelines.

Table 1.1 Summary of publication outlets of the papers in this thesis, including peer-review process history

Publication Title	Target Journal	Peer Review	Impact Factor / SJR	Submission Date	Acceptance Date	# Re-visions
Bibliometric Study on the Use of Machine Learning as Resolution Technique for Facility Layout Problems	IEEE Access	single-anonymised	3.4 / 0.849 (Q1)	January 11, 2021	January 20, 2021	2
gym-flp: A Python Package for Training Reinforcement Learning Algorithms on Facility Layout Problems	Operations Research Forum	single-anonymised	*Not reported* / 0.403 (Q2)	March 27, 2021	January 29, 2024	3

(continued)

Table 1.1 (continued)

Publication Title	Target Journal	Peer Review	Impact Factor / SJR	Submission Date	Acceptance Date	# Re-visions
Deep reinforcement learning for layout planning—An MDP-based approach for the facility layout problem	Manufacturing Letters	single-anonymised	1.9 / 0.551(Q2)	March 9, 2023	September 17, 2023	1
From Theory to Application: Investigating the Generalizability of Facility Layout Problems Using a Deep Reinforcement Learning Approach	Production Engineering Research and Development	single-anonymised	1.7 / 0.540 (Q2)	January 12, 2025	May 6, 2025	2

Table 1.2 Injection of the research design into the DSR guidelines (Hevner et al., 2004)

Guideline	Description	Operationalisation
1: Design as an artefact	Design Science Research must provide a viable artefact in the form of a construct, model, method or instantiation	This research establishes three main elements of artefacts: An MDP formalism for the FLP in RL settings (Artefact type: model) The simulation environment and training procedure (Artefact types: construct and instantiation) A Generalisation study of applying the above method to various instances of the problem domain (Artefact type: method, model)
2: Problem Relevance	The objective of design-science research is to develop technology-based solutions to important and relevant business problems	The relevance of tackling the FLP using RL tools was motivated by the experiences of the head researcher in factory planning consulting projects. The amount of interest and research on solving FLPs, as evidenced by the systematic literature review herein, testifies to its relevance.

(continued)

Table 1.2 (continued)

Guideline	Description	Operationalisation
3: Design Evaluation	The utility, quality and efficacy of a design artefact must be rigorously demonstrated via well-executed evaluation methods	Formal proof methods are used to validate the effectiveness of the designed mechanisms. At the code and module levels, this includes unit and acceptance tests. On the procedural level, the experiment outcomes are tested against the optimisation criterion of layout improvement, which is defined as a reduction in MHC.
4: Research Contributions	Effective design-science research must provide transparent and verifiable contributions in the areas of the design artefact, design foundations, and/or design methodologies	The research has a broad impact on the application domain, utilising previously conceived problem sets and providing open-source artefacts via a standard coding outlet to facilitate cross-checking and further development by other researchers.
5: Research Rigour	Design-science research relies upon the application of rigorous methods in both the construction and evaluation of the design artefact	This work is based on past research in the fields of Reinforcement Learning (algorithms, simulation backbones, and evaluation methods) and facility layout planning (a formalised model of MHC). Furthermore, rigorous testing techniques are employed to ensure the efficacy of the artefacts at the code level.

(continued)

Table 1.2 (continued)

Guideline	Description	Operationalisation
6: Design as a Search Process	The search for an effective artefact requires utilising available means to reach desired ends while satisfying laws in the problem environment	The usefulness of the underlying MDP has been iteratively tested by altering design aspects (state representation, action space, and reward structure) and observing the resulting evaluation. Furthermore, the guiding laws and constraints of the domain environment (material handling cost improvement, spatial plant constraints, layout feasibility) have been decomposed into sub-problems and solved sequentially in a divide-and-conquer manner.
7: Communication of Research	Design-science research must be presented effectively both to technology-oriented and management-oriented audiences	Four substantial reports on the results of the individual artefacts have been published for a wider scientific audience, both in more operations research-focused (method) and managerial outlets (construct, instantiation). This dissertation work further compiles all publications into a single document and discusses the results synergistically for the problem domain.

Theoretical Background for Automated Layout Planning Using Reinforcement Learning

2

To facilitate a comprehensive evaluation of the novelty and relevance of the four publications in this cumulative dissertation, the following sections present a detailed overview of the knowledge bases for Facility Layout Problems and Reinforcement Learning. The elements described in Sect. 2.1 are essential for understanding the design requirements for solving FLPs. The subsequent Sect. 2.2 then explores the solution dimension for the thesis, which is vital for translating FLP concepts into the RL domain.

2.1 Facility Layout Planning

The Factory Layout Problem (FLP) is a significant area in operations and production management concerned with the most efficient physical arrangement of departments, work areas, equipment, and storage spaces within a manufacturing facility (Pérez-Gosende et al., 2021) based on one or more objectives and in accordance with various constraints (Heragu and Kakuturi, 1997). At its core, FLPs aim to reduce material handling costs, optimise workflows, and ensure safety and ergonomic conditions (Tompkins et al., 2010). In modern manufacturing settings, where product mixes change rapidly and global competition demands higher efficiency, the design and continuous improvement of facility layouts can deliver a substantial competitive advantage (Drira et al., 2007).

B. Heinbach, *Reinforcement Learning-Based Planning of Factory Layouts*, Findings from Production Management Research ,
https://doi.org/10.1007/978-3-658-51554-6_2

From a production management perspective, a practical facility layout impacts throughput times, work-in-process (WIP) levels, lead times, and overall resource utilisation (Muther, 1973). Consequently, it becomes crucial in ensuring a smooth flow of materials, timely production schedules, and consistent product quality (Kulturel-Konak, 2012). Over the past several decades, a broad spectrum of modelling approaches (ranging from mathematical formulations to AI-driven frameworks) and solution methodologies (exact, heuristic, meta-heuristic, and intelligent) have been developed (Xiao et al., 2017). These advancements reflect the complexities of contemporary manufacturing systems, where facility layouts must adapt to dynamic demand, integrate new process technologies, and meet evolving sustainability standards (Zhu et al., 2022).

In the literature, material handling costs (MHC) are often used as a key quantitative metric for assessing the efficiency of a factory layout (Emami and S. Nookabadi, 2013). Other relevant criteria can be of a technical, social or economic nature (Pawellek, 2014) such as area efficiency (Raman et al., 2009), changeability (Wiendahl et al., 2024), ergonomic factors like noise (Adem, 2023), energy requirements (Allen-Zhu et al., 2016), or climate-related considerations (Dombrowski and Marx, 2018). The FLP can be modelled as a single-criterion or multi-criteria optimisation problem (Singh and Sharma, 2006).

According to the analysis by Hosseini-Nasab et al. (2018), current research on solution approaches in FLP predominantly focuses on the aforementioned population-based metaheuristics. These techniques have long been regarded as effective methods for multi-objective optimisation. However, due to their substantial requirements for large populations and numerous iterations, they exhibit poor scalability when applied to huge problem sets (Li et al., 2024; Lust and Teghem, 2010).

Figure 2.1 above shows a taxonomy of the facility layout problem mapped by Hosseini-Nasab et al. (2018). Since it will be an integral part of this thesis to design artefacts that match the respective FLP components, the following subsections will focus on discussing the elements from the taxonomy with the most relevance to this work.

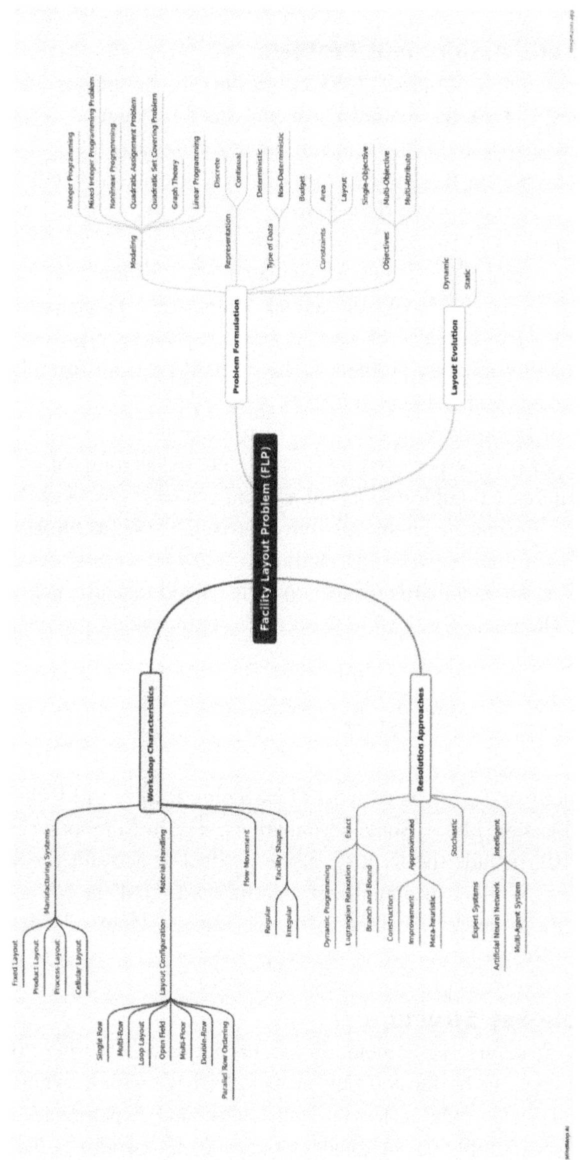

Fig. 2.1 Classification of facility layout problem (Bouramtane et al., 2024) adapted from Hosseini-Nasab et al. (2018)

2.1.1 Problem Formulation

2.1.1.1 Quadratic Assignment Problem

The Quadratic Assignment Problem (QAP) goes back to Koopmans and Beckmann (1957). The plant site is divided into equal-sized rectangular blocks with the same area and shape, and one facility is assigned to each block iteratively until an optimal assignment is identified in terms of low material handling costs (Simmons, 1969). The name QAP derives from the objective function, which is a second-degree function of the variables (Kusiak and Heragu, 1987). A discrete FLP with unequal-sized facilities can be formulated as a Quadratic Set Covering Problem (QSP) (Bazaraa, 1975). As the actual content of a block is irrelevant for the solution, the QAP can also be used to solve problems outside of a plant context, such as facility location problems. The original problem formulation of the QAP by Koopmans and Beckmann (1957) is as follows:

Given is a set of facilities $N = \{1, 2, \ldots, n\}$ to be assigned to a set of locations of the same size N. A permutation p represents the non-repeating unique assignment of facilities 1 though n to locations. Thus, $S_n = p : N \rightarrow N$ is the set of all permutations. Further given are the flow matrix $F = (fij)$ as an nxn matrix where fij represents a flow between the facilities i and j as well as the distance matrix $D = (dij)$ as an nxn matrix where dij is the fixed distance between the locations i and j. The goal of a QAP is to solve the optimisation problem

$$\min_{p \in S_n} \sum_{i=1}^{n} \sum_{j=1}^{n} f_{ij} * d_{p(i)p(j)} \tag{2.1}$$

where each product $f_{ij} * d_{p(i)p(j)}$ is the cost of assigning facility i to location $p(i)$ and facility j to location $p(j)$. . Some formulations, e.g. in Simmons (1969) or Scholz et al. (2010), include the term c_{ij} in the product to account for different cost per unit flows. While conceptually straightforward, QAP is NP-hard, and exact solution methods quickly become infeasible for $n > 15$ or so (Meller and Gau, 1996).

2.1.1.2 Flexible Bay Structure

The Flexible Bay Structure (FBS) notation has been created by Tong (1991). It allows the departments to be located only in parallel bays with varying widths. Bays are bounded by straight aisles on both sides, and departments are not allowed to span over multiple bays (Konak et al., 2006). The width of the bays is determined by the cumulated area demand of the facilities assigned to each one, using the equation below (Ulutas and Kulturel-Konak, 2012):

$$\text{Length of bay } i = \frac{\text{Sum of department areas in bay } i}{\text{Width of the facility}} \tag{2.2}$$

Layouts are generated by manipulating two vectors: the permutation p, indicating the order in which facilities are located, and the bay breaks b. This is a binary vector that represents the last facility in a bay and is of the same length as the permutation. The interpretation of this vector varies throughout the literature: while Garcia-Hernandez et al. (2020) indicate a bay division with the value 1, Ulutas and Kulturel-Konak (2012) represent break positions with a 0. This means that the last position in b is always equal to 1. For instance, in Fig. 2.2, the bay break vector would be $b = \{0, 0, 1, 0, 0, 1, 0, 0, 1\}$. For clarity, the positions in the permutation and the bay breaks they match are highlighted in bold in the figure.

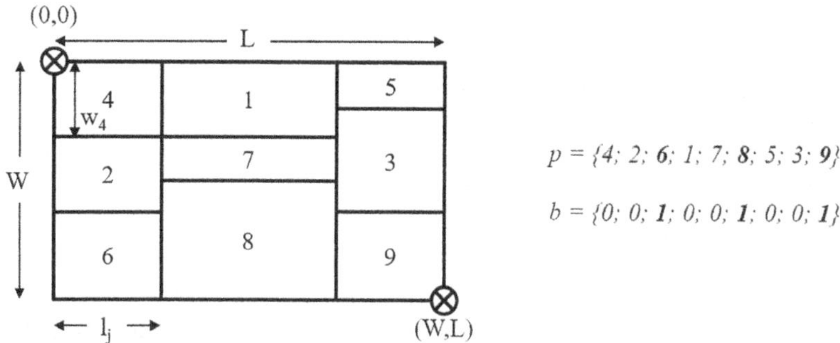

Fig. 2.2 Example of a layout of size $n = 9$ created using the FBS notation

2.1.1.3 Slicing Tree Structure

The Slicing Tree Structure (STS) goes back to Tam (1992). Friedrich et al. (2018) and Ripon et al. (2013) use a layout encoding that encompasses three distinct items: a permutation vector p, a slicing order vector s, and an orientation vector o. While p is of size $i = n$, o and s are of size $j = n - 1$. p has the same meaning as in QAP and FBS. s describes the position of a cut within the layout, and o denotes the direction of the cut, which can be horizontal or vertical. This is the key difference compared to FBS, as STS allows cuts in two directions rather than one.

Given the plant dimensions W, L and all area requirements a_i for the facilities, the layout creation follows the following procedure (Scholz, 2010):

1. Build a slicing tree by sequentially cutting p at the positions indicated by s in direction o and assign the thus created new vectors to the sub-layouts. For any vector slice of size 1, i.e. holding only one facility, it is assigned as an outer node (tree leaf).
2. Starting from the leaves in the lowest tree level, the area requirement of their parent node is computed. This is repeated in a bottom-up manner until the root node is reached.
3. The empty layout will now be sliced along the order as given by s. . The slicing position is determined similarly to the FBS equation above, i.e. by dividing the area required in one sub-layout by the width or length of the plant (depending on the slicing direction).

Figure 2.3 shows an illustrative example of the STS approach. The layout in Fig. 2.3b is encoded by the permutation $p = \{4, 2, 1, 6, 7, 8, 5, 3, 9\}$, the slicing order $s = \{4, 5, 3, 2, 8, 1, 6, 7\}$ and the slicing orientations $o = \{0, 1, 1, 0, 0, 1, 1, 0\}$ leading to the slicing tree in Fig. 2.3a. After the first cut, located in the left-hand child node after the root, the layout is split into the sub-layouts with the permutation slices $p_{1,l} = \{4, 2, 1, 6\}$ and $p_{1,r} = \{7, 8, 5, 3, 9\}$. The index of the superscript indicates the number of cuts performed in the range $1 \ldots n - 1$, and the letter indicates the side of the new sub-layout. Given a vertical cut, the width of the sub-layout corresponds to W whereas the length can be computed as $l_{1,l} = \frac{a4+a2+a1+a6}{W}$. This computation is repeated recursively until the leaf nodes are reached. For a more detailed explanation of the STS approach, the reader shall be referred to Friedrich et al. (2018).

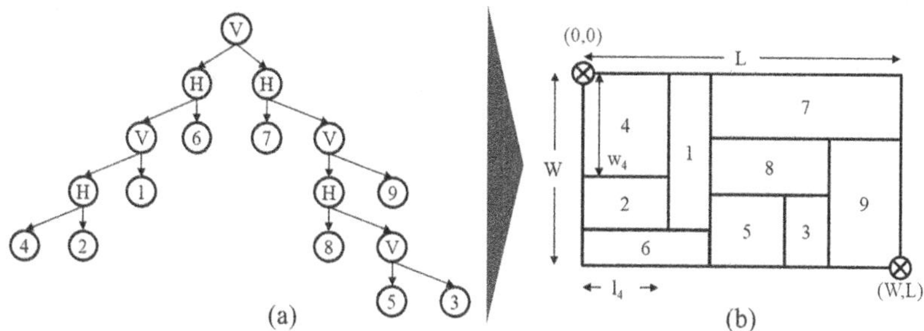

Fig. 2.3 Example for a graphical representation of slicing tree and resulting layout ($n = 9$)

2.1.1.4 Open Field Layout Problem

The open field problem (OFLP) is based on the collision-free unequal area FLP mixed-integer linear programming (MILP) solution by Montreuil et al. (1993). From a practice-based factory planning perspective, this may be the closest resemblance to a real-world planning approach. The OFLP is characterised by the size variables of the individual facilities and the plant itself, i.e. W, L, w and l. This approach does not rely on divisions being made; facilities can move around freely within the plant, constrained only by the plant's dimensions, that is, in its most basic representation. To build layouts, no encoding mechanisms are required other than the facility dimensions or areas and their coordinates. The freedom from collisions with structures or other facilities, which is inherent in the previous modelling techniques, needs to be included mathematically or programmatically in the solution method for this type of layout representation. An example of an OFLP problem is given in Fig. 2.4.

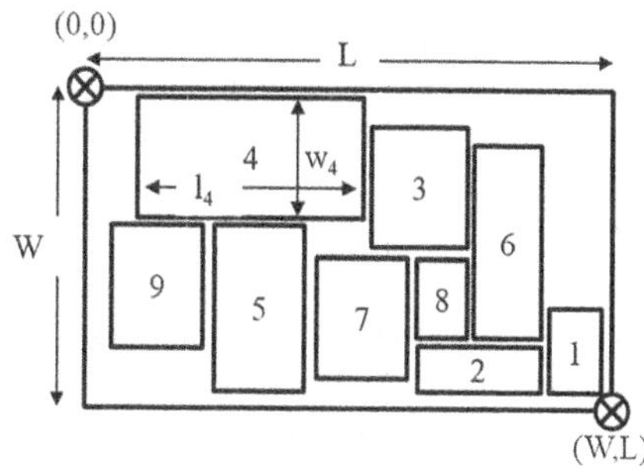

Fig. 2.4 Graphical representation of a possible solution for the OFLP environment ($n = 9$)

2.1.2 Layout Evolution

Plant layout problems are typically divided into two categories: Static Facility Layout Problems (SFLP) and Dynamic Facility Layout Problem (DFLP) (Burggräf, Adlon et al., 2021; Rosenblatt, 1986), depending on the duration of the planning horizon. A static facility layout is designed for a single planning horizon under the assumption that production requirements (e.g., product demand, mix) remain relatively stable (Singh and Sharma, 2006). Traditional studies tended to treat FLP as static, producing a one-shot solution that remains in place for years (Meller and Gau, 1996). While simpler to model and implement, static layouts can become suboptimal in volatile markets, leading to increased material handling costs and bottlenecks over time (Pérez-Gosende et al., 2021).

However, in practice, material flows are unlikely to remain static over extended planning periods. Factors such as intense global competition, rapid technological advancements, and shorter product life cycles compel companies to reassess and adjust their facility layouts regularly (Balakrishnan et al., 1992). Nicol and Hollier (1983) support the necessity of dynamic treatment of the problem by highlighting the frequency of occurrence of drastic layout changes. As a result, dynamic layouts have gained prominence alongside static layouts (Jolai et al., 2012). A dynamic layout acknowledges that manufacturing environments are constantly evolving. The DFLP divides the planning horizon into discrete periods, each with potentially different flow intensities, department requirements, or production volumes (Balakrishnan and Cheng, 2000). By optimising both the initial arrangement and the subsequent re-layouts (or reconfiguration decisions), DFLP aims to minimise the sum of material handling plus rearrangement costs across all periods (Xiao et al., 2017).

Empirical results from various industries, particularly footwear (Ulutas and Islier, 2015) and textile (Xiao et al., 2017), show that dynamic planning often outperforms a static approach in scenarios where demand or product design changes regularly. However, reconfiguration also entails production downtime and capital expenditure, so the benefits must outweigh the rearrangement costs (Balakrishnan and Cheng, 2000).

2.1.3 Resolution Approaches

2.1.3.1 Exact Methods

Significant variations exist in the solution approaches to such problems. In the early stages of FLP research, the focus was on obtaining optimal solutions for the QAP (Kusiak and Heragu, 1987). These tended to be solved with exact methods (e.g., branch-and-bound, branch-and-cut, mixed-integer linear programming), which seek provably optimal solutions (Krarup and Pruzan, 1983). However, since both QAP and QSP have been validated as NP-complete for problem sets exceeding 15 production units (Kusiak and Heragu, 1987), exact methods become ineffective for problem sizes involving 15 (Meller and Gau, 1996) or 18 production units (Chraibi et al., 2019).

While theoretically attractive, even workshops with just a few hundred square meters of workspace often exceed 20 production resources in their layouts, rendering exact methods viable only for minor problems due to exponential growth in the solution space (Meller and Gau, 1996). Nonetheless, exact approaches play a vital role in benchmarking approximate methods and solving subproblems or partial layout optimisations.

2.1.3.2 Heuristic Methods

To overcome the limitations of computational complexity in FLPs, heuristic methods were conceived early on. Heuristics are problem-specific or rule-based techniques aimed at quickly generating "good enough" solutions (Muther, 1973). They generally do not guarantee global optimality but are computationally efficient and straightforward to apply in industrial contexts (Tompkins et al., 2010). Examples include:

- ALDEP (Automated Layout Design Program)—builds layouts incrementally based on closeness ratings (Seehof and Evans, 1967)
- CORELAP (Computerised Relationship Layout Planning)—constructs layouts by placing departments with the highest "relationship" first (Lee and Moore, 1967), similar to ALDEP
- CRAFT (Computerised Relative Allocation of Facilities Technique)—an improvement heuristic that starts with a layout, then iteratively swaps pairs of departments to reduce cost (Armour and Buffa, 1963)

2.1.3.3 Meta-Heuristic Approaches

Meta-heuristics guide local or constructive heuristics with overarching strategies for global exploration. Widely applied to FLP, they excel at escaping local optima and can handle multi-period or multi-objective variants. The landscape of meta-heuristic algorithms for FLP is vast in breadth and depth. Interested readers shall therefore be directed to these review articles: García-Hernández et al. (2014), Hosseini-Nasab et al. (2018), Burggräf, Adlon et al. (2021) and Chemim et al. (2021). Below follows a brief overview of popular meta-heuristics:

- Simulated Annealing (SA)—mimics physical annealing, occasionally accepting worse solutions to avoid local minima (cf. (Ku et al., 2011)).
- Genetic Algorithms (GA)—uses evolutionary ideas (selection, crossover, mutation) to "breed" better layouts across generations (cf. (Islier, 1998)).
- Tabu Search (TS)—systematically forbids or penalises revisiting recently explored solutions to encourage broader exploration (cf. (Kulturel-Konak, 2012)).
- Particle Swarm Optimisation (PSO)—treats solution updates as movements of particles influenced by individual and collective best experiences (cf. (Paul et al., 2006)).
- Ant Colony Optimisation (ACO)—inspired by the foraging behaviour of ants. It utilises a population of artificial ants that construct solutions based on pheromone trails, which represent the quality of these solutions (cf. (Kulturel-Konak and Konak, 2011))

In the context of both meta-heuristic and heuristic methods, it is crucial to distinguish between Construction and Improvement algorithms. Construction algorithms begin on a blank canvas and assemble a layout by placing departments incrementally, typically by using a relationship diagram. Examples are ALDEP and CORELAP. While they deliver fast results and can incorporate qualitative preferences, the final solution largely depends on the order in which facilities are arranged. Improvement algorithms, such as CRAFT, take an existing layout and apply local search or iterative changes to enhance it, for instance, through pairwise swapping. While these algorithms can potentially improve layout quality significantly, the final result often depends on the initial layout in a way that they encounter local rather than global optima (Kim and Kim, 1995). Meanwhile, meta-heuristics often internalise both construction-like initialisation (random or heuristic-based) and iterative improvement (mutation, local search) within their frameworks (Balakrishnan et al., 2003) to mitigate exactly these adverse effects.

2.1.3.4 Intelligent Approaches

In recent decades, intelligent approaches to FLP have garnered increasing attention (Ulutas and Kulturel-Konak, 2012). These methods extend beyond classical optimisation or meta-heuristics by integrating Artificial Intelligence (AI) and expert system paradigms. Often, they aim to handle qualitative criteria, uncertainty, and non-quantifiable preferences that conventional methods may struggle to incorporate. Below are some primary intelligent techniques used in FLP.

Expert systems—also known as knowledge-based systems—encode the tacit knowledge of experienced facility planners into a rule-based decision engine

(García-Hernández et al., 2014; Shouman et al., 2001). The advantage of expert systems lies in substituting human expert knowledge to avoid burdening the expert (García-Hernández et al., 2015). However, maintenance can become an issue if the knowledge base grows large or if the facility's operations change frequently, necessitating regular updates to the rules.

Artificial Neural Networks (ANNs) are computational models inspired by the human brain, consisting of interconnected nodes (neurons) that process information. They are particularly useful for pattern recognition, classification, and regression tasks (LeCun et al., 2015). Especially in their earlier forms, such as self-organising maps (SOM) or Hopfield networks, have been applied to FLP to identify clusters, manage adjacency, or perform pattern-based allocations. Key ways ANNs can support FLP are manifold. For instance, in Unsupervised Learning paradigms, they can perform clustering and grouping, learning patterns in departmental inter-flow data by grouping departments with high mutual flows into clusters (Potocnik et al., 2014). With Supervised Learning, in turn, they can be used as regressors for approximation or cost estimation to predict travel times or flow intensities from historic data (c.f. (Niebles et al., 2016; Tam and Tong, 2003)). Alternatively, when trained as classifiers, ANNs can be used as decision support systems to assist in the layout evaluation procedure by surrogating expert ratings (Garcia et al., 2018). However, especially for the latter two cases, practical training of ANNs requires a substantial amount of historical data related to facility layouts, which can be scarce (Hu and Yang, 2019).

2.1.3.5 Manual Planning Procedures

The resolution approaches introduced up to this point have been, by and large, of computational and algorithmic nature. While harnessing computing power to optimise layouts based on quantifiable criteria, FLPs frequently involve multiple, competing objectives, such as minimising cost, enhancing safety, maximising the proximity of related departments, and improving environmental or ergonomic factors. To address these multi-criteria or multi-attribute aspects, layout planning often turns to structured decision-support methods. Among these, the Analytic Hierarchy Process (AHP), Analytic Network Process (ANP), and Technique for Order of Preference by Similarity to Ideal Solution (TOPSIS) are widely applied.

AHP decomposes a complex decision problem into a hierarchical structure of goals, criteria, sub-criteria, and alternatives. Decision-makers pairwise compare elements at each level and use a consistent scale to quantify judgments. This way, AHP structures the evaluation of criteria such as material handling cost, safety requirements, department adjacency, and environmental impact in group decision-making (Saaty, 1980, 1990). While the concept is straightforward, the process assumes independence between criteria to avoid introducing bias towards any specific criterion. Furthermore, the process is prone to over- or underestimation of certain aspects, especially in cross-departmental group decision-making processes.

ANP extends AHP by allowing interdependencies and feedback among criteria and sub-criteria. Instead of a strict top-down hierarchy, ANP models the decision as a network, capturing the reality that criteria can influence each other. ANP handles the aforementioned mutual dependencies, making it valuable for multi-period or complex FLPs where departmental proximity may affect multiple objectives simultaneously. ANP is more realistic than a purely hierarchical approach and captures feedback loops, but requires more expertise to implement correctly (Saaty, 2013).

TOPSIS is based on the idea that the best alternative is the one closest to the ideal solution and farthest from the negative-ideal solution in a multi-dimensional attribute space. After identifying layout alternatives (e.g., using heuristics, metaheuristics, or simulation), each layout is evaluated against multiple criteria, such as distance metrics, safety scores, or adjacency preferences. TOPSIS then ranks these layouts by computing their relative closeness to an ideal point. TOPSIS is straightforward computationally, but requires normalised data and weighting of criteria, which, similar to AHP, can introduce subjectivity if stakeholder consensus on weights is lacking (Hwang, 1981).

2.2 Reinforcement Learning

Reinforcement Learning (RL) constitutes a foundational paradigm in Artificial Intelligence whereby an autonomous agent learns to perform sequential decision-making through interaction with an environment. Formally, the agent observes states, selects actions, and receives rewards that quantify the quality of its decisions. The overarching objective of an RL agent is to discover an optimal behaviour (policy) that maximises the expected cumulative reward (or return) over time. RL provides a flexible framework for handling problems that exhibit uncertainty, high-dimensional state spaces, and complex dynamics—attributes commonly encountered in industrial settings such as facility layout optimisation (Sutton and Barto, 2018).

Despite the rich tradition of mathematical programming and heuristic approaches to facility layout problems, contemporary research highlights the potential of RL-based solutions to address these combinatorial design challenges in a more adaptive and data-driven fashion. This section provides an in-depth overview of RL, from fundamental concepts to leading-edge algorithms, and reviews their application and effectiveness in facility layout optimisation tasks.

2.2.1 Mathematical Foundations of Reinforcement Learning

2.2.1.1 Markov Decision Process

The predominant formalism underlying Reinforcement Learning is the Markov Decision Process (MDP), defined by the tuple $(\mathcal{S}, \mathcal{A}, \mathcal{P}, \mathcal{R})$. At each time step t, the system resides in state $s_t \in \mathcal{S}$, the agent executes an action $a_t \in \mathcal{A}$, and the system transitions to a new state $s_{t+1} \sim \mathcal{P}(\cdot|s_t, a_t)$, yielding a reward $r_{t+1} = \mathcal{R}(s_t, a_t)$. A policy π prescribes the agent's behaviour in a stochastic manner: $\pi(a|s)$ (Sutton and Barto, 2018). The agent's objective is to find π that maximizes its expected discounted return $\mathbb{E}_{\tau\sim\pi_\theta}\left[\sum_{t=0}^{\infty} \gamma^t r_t\right]$.

2.2.1.2 Bellmann Equations

Central to MDP theory are the Bellman equations, which express value functions in terms of one-step rewards and the discounted value of successor states. For a given policy π, , the Bellman expectation equation (Bellman, 1957) for the action-value function Q^π is

$$Q^\pi(s, a) = \mathbb{E}\left[r + \gamma \max_{a'} Q^\pi(s', a')|s, a, a' \sim \pi\right]. \tag{2.3}$$

When seeking optimality, Q-learning and other dynamic programming approaches utilise the Bellman optimality equation:

$$Q^*(s, a) = \mathbb{E}\left[r + \gamma \max_{a'} Q^*(s', a')|s, a\right], \tag{2.4}$$

where r is the immediate reward, $\gamma \in [0, 1)$ is the discount factor, and the expectation is taken concerning the environment's stochastic transitions. By iterating Bellman backups or equivalent sample-based updates (e.g., Temporal-Difference (TD) learning), one converges to the optimal solution Q^* (Dong et al., 2020; Sutton and Barto, 2018).

2.2.1.3 Optimisation Strategies

In practice, RL algorithms solve MDPs through either of the following:

1. Direct Dynamic Programming: Feasible in small state spaces when the transition model $\mathcal{P}$ is fully known (e.g., value iteration, policy iteration).
2. Model-Free Approximation: Learning from experience via TD or Monte Carlo methods.
3. Model-Based Planning: Learning or using a known model and performing simulated rollouts, as in Dyna-Q (Sutton and Barto, 2018) or more advanced model-based RL algorithms (Ha and Schmidhuber, 2018).

When function approximation (e.g., neural networks) is introduced, the problem translates to an iterative gradient-based optimisation, either aiming to fit the Bellman target for value-based methods or to perform gradient ascent on the policy's return in policy-based methods (Schulman et al., 2017). Actor-critic architectures effectively blend these two approaches.

2.2.2 Model-Free vs. Model-Based Reinforcement Learning

2.2.2.1 Model-Free Reinforcement Learning

In model-free RL, the agent does not rely on an explicit model of the environment's dynamics, denoted typically by the transition probability function $P(s'|s, a)$. Instead, it directly learns to estimate value functions or policies from sampled transitions, observing states, actions, rewards, and next states (s, a, r, s') through repeated interaction (Sutton and Barto, 2018). Q-learning (Watkins and Dayan, 1992) and SARSA (Rummery and Niranjan, 1994) are canonical examples of model-free RL algorithms, each iteratively refining an estimate of the action-value function via TD updates. Advantages of model-free RL include conceptual simplicity and broad applicability, at the cost of potentially higher sample complexity since the agent cannot "plan" future steps based on a learned or known model (Sutton, 1990; Sutton and Barto, 2018).

2.2.2.2 Model-Based Reinforcement Learning

Contrastingly, model-based RL presupposes, or actively learns, a model of the environment's transition and reward dynamics. This model is then used for planning, meaning the agent can predict future trajectories and evaluate alternative action sequences without directly executing them in the actual environment

(Deisenroth, 2011). By leveraging simulated rollouts or methods such as Monte Carlo Tree Search (MCTS), model-based approaches often achieve superior sample efficiency, since fewer real interactions are required to converge to a high-performance policy (Ha and Schmidhuber, 2018). However, errors in the learned model can lead to suboptimal or biased decisions. Recent breakthroughs, including AlphaZero (Silver et al., 2016) and MuZero (Schrittwieser et al., 2020), highlight the efficacy of combining model-based planning with deep neural networks for both representation learning and policy improvement.

2.2.3 Value-Based vs. Policy-Based Methods

2.2.3.1 Value-Based Methods

Value-based RL is characterised by the estimation of a value function that indicates the expected return from a given state-action pair under some policy. In particular, Q-learning endeavours to learn the optimal action-value function $Q^*(s, a)$ that satisfies the Bellman optimality equation above: Upon convergence, the Q-learning agent derives its policy by greedily selecting actions that maximise $Q^*(s, a)$. . Algorithms of this form typically excel in discrete or moderate-sized state-action spaces and can reuse off-policy data, improving sample efficiency (Lin, 1992). However, in high-dimensional continuous action spaces, purely value-based methods often struggle due to the need to discretise or approximate Q-values over large continuous domains (Silver et al., 2014).

2.2.3.2 Policy-Based Methods

Policy-based RL approaches, conversely, aim to learn a parameterised policy $\pi_\theta(a|s)$ directly, bypassing the need to represent a value for each state-action pair (Williams, 1992). The agent optimises the expected return

$$J(\theta) = \mathbb{E}_{\tau \sim \pi_\theta}\left[\sum_{t=0}^{\infty} \gamma^t r_t\right], \tag{2.5}$$

where τ denotes trajectories generated by the policy π_θ. . Gradient-based optimisation is commonly employed, with the policy gradient theorem (Sutton et al., 1999) providing a foundation for algorithms such as REINFORCE (Sutton et al., 1999) and Asynchronous Advantage Actor-Critic (A3C) (Mnih et al., 2016). While policy-based methods often exhibit improved stability and handle continuous action spaces naturally, they may suffer from high variance in gradient estimates and typically require on-policy data collection. Techniques like actor-critic and trust-region methods (e.g., TRPO, PPO) have been introduced to mitigate these issues and improve training stability (Schulman et al., 2017).

2.2.4 On-Policy vs. Off-Policy Learning

On-policy learning refers to the agent learning and improving the same policy it uses to act in the environment. In contrast, off-policy methods can learn an optimal target policy while using a different behaviour policy to gather data. For example, SARSA learns the value of the current policy, whereas Q-learning learns the value of the optimal policy even while following an exploratory behaviour.

Off-policy algorithms are generally more sample-efficient because they can reuse past experiences for learning. On-policy methods must sample fresh trajectories after each policy update, so they cannot fully leverage past data. This means on-policy approaches typically require more interactions with the environment to reach a given level of performance compared to off-policy methods.

On-policy approaches tend to be more stable and have convergence guarantees under certain conditions, since they update the policy using actual actions (avoiding drastic jumps). Off-policy methods can converge faster to a high-performing policy by directly learning from optimal value estimates; however, they risk instability or divergence if not carefully controlled. In practice, off-policy algorithms often require extra techniques (like target networks or double Q-learning) to ensure stable learning.

2.2.5 Common Reinforcement Learning Algorithms

2.2.5.1 Q-Learning

Formalised by Watkins and Dayan (1992), Q-learning remains one of the most influential model-free, off-policy RL algorithms. Its core update rule is:

$$Q(s, a) \leftarrow Q(s, a) + \alpha\left[r + \gamma \max_{a'} Q(s', a') - Q(s, a)\right], \tag{2.6}$$

where α is the learning rate, γ is the discount factor, and the agent updates Q values toward the Bellman optimality target $r + \gamma \max_{a'} Q(s', a')$. Q-learning is proven to converge to the optimal action-value function Q^* for finite MDPs, provided it sufficiently explores all state-action pairs and uses an appropriate decay for α (Sutton and Barto, 2018; Watkins and Dayan, 1992). This off-policy nature permits the agent to learn about an optimal greedy policy while following an exploratory behaviour in practice.

Q-Learning has been applied in various industrial domains for decision-making and control tasks. Zahedi-Seresht et al. (2024) used a Q-learning approach to optimise oil well settings, improving the oil recovery factor. Their case study showed that Q-learning could iteratively discover efficient initial production rates in a simulated reservoir, aiding engineers in maximising output. Zhang et al. (2024) developed a multi-AGV (Automated Guided Vehicle) route planning method based on a shortest-time Q-learning algorithm for smart warehouses. The classical Q-learning-based controller was enhanced to handle the complex "chessboard-like" layout of an automated warehouse, significantly improving route planning efficiency and reducing travel time for multiple AGVs operating concurrently.

2.2.5.2 SARSA

SARSA (State-Action-Reward-State-Action) is an on-policy variant of temporal-difference learning (Dong et al., 2020). Its update uses the agent's actual next action a' under the current behaviour policy (often $\in$-greedy) instead of the $max_{a'}$ operator:

$$Q(s, a) \leftarrow Q(s, a) + \alpha\left[r + \gamma Q\left(s', a'\right) - Q(s, a)\right] \tag{2.7}$$

Consequently, SARSA converges to the Q-values of the policy it follows, incorporating the effects of any exploratory strategy (e.g., $\in$-greedy). In specific environments (e.g., cliff-walking), SARSA can learn more conservative, "safer" paths by accounting for the possibility of dangerous exploratory actions (Sutton and Barto, 2018). However, these conservative policies can quickly lead to poor performance in scenarios with larger state or action spaces (Momenikorbekandi and Abbod, 2023). Nonetheless, the SARSA algorithm has been demonstrated to be effective, e.g. in AGV path planning (Liao et al., 2020) and scheduling (Aissani et al., 2008).

2.2.5.3 Deep Q-Networks

While Q-learning and SARSA are straightforward in tabular or small-scale problems, they face significant challenges when the state space is large or continuous.

Deep Q-Networks (DQN), pioneered by Mnih et al. (2015), address these limitations by approximating the Q-function with a deep neural network, enabling the agent to operate directly on high-dimensional inputs such as raw images.

Two key innovations in DQN stabilise training:

1. Experience Replay: The agent stores transitions (s, a, r, s') in a replay buffer and randomly samples mini-batches for training. This breaks the correlation among consecutive samples and improves data efficiency.
2. Target Networks: A slowly updated "target" network is used to compute the Bellman backup target, decoupling rapidly changing Q estimates from the bootstrap target.

These modifications reduce non-stationarity and catastrophic forgetting in deep neural network training. Compared to vanilla Q-learning, DQN enables agents to handle vast or unstructured observation spaces, thereby substantially broadening the applicability of RL. DQN catalysed a wave of Deep Reinforcement Learning research, with subsequent enhancements such as Double DQN, Prioritised Replay, Duelling Networks, and Rainbow DQN further improving stability and learning speed.

DQN has seen successful industrial applications, especially where state spaces are large or complex. Waschneck et al. (2018) applied DeepMind's DQN algorithm to optimise semiconductor production scheduling. In their real-world fab case, a DQN agent learned dispatching policies that outperformed traditional heuristic rules, effectively reducing late jobs in a complex job-shop environment. Sumanas et al. (2022) reported using a deep Q-learning method to improve the precision of an industrial robot arm, where a DQN-based controller was trained to self-correct positioning errors in real-time, resulting in a significant increase in the robot's positioning accuracy and repeatability after only a few hundred training iterations.

2.2.5.4 Trust Region Policy Optimisation

Trust Region Policy Optimisation (TRPO) is a foundational policy-based or actor-critic algorithm introduced by Schulman et al. (2015). TRPO addresses instability in policy gradient methods by enforcing a trust region constraint that bounds the change in the policy at each update:

$$\max_{\theta} \mathbb{E}_{s \sim \pi_{\theta_{old}}} \left[\frac{\pi_{\theta}(a|s)}{\pi_{\theta_{old}}(a|s)} A^{\pi_{\theta_{old}}}(s, a) \right] \text{subject to } D_{KL}\left[\pi_{\theta_{old}}(\cdot|s)| \ |\pi_{\theta}(\cdot|s)\right] \leq \delta \tag{2.8}$$

where $A^{\pi_{\theta_{old}}}(s,a)$ is an advantage function, D_{KL} is the Kullback–Leibler divergence measuring policy dissimilarity, and δ is a trust-region threshold. By constraining the new policy π_θ to remain close to the old policy $\pi_{\theta_{old}}$, , TRPO ensures more stable improvement and avoids destructive large policy updates. Despite its strong theoretical underpinnings and empirical success, TRPO involves solving a relatively expensive constrained optimisation subproblem at each iteration. Subsequent algorithms like Proximal Policy Optimisation (Schulman et al., 2017) sought to simplify this approach while retaining many of its performance benefits.

TRPO has been explored in multi-agent energy optimisation. For example, Thattai et al. (2023) present a home energy management system that uses a multi-agent TRPO-based RL approach to optimally schedule household appliances and battery storage. Their method aimed to minimise energy cost and maintain user comfort and was tested with real load data. Furthermore, from a manufacturing perspective, TRPO has been tested in job-shop scheduling: Kuhnle et al. (2021) employ TRPO for adaptive production control in job-shop manufacturing. TRPO proves robust and superior to traditional scheduling heuristics, optimising machine utilisation and order waiting times through Reinforcement Learning (RL). The study finds that reward design has a significant impact on performance, with multi-criteria optimisation balancing trade-offs. Fixed-action episodes enhance learning speed, while direct action mapping ensures faster convergence. Ultimately, RL-based scheduling outperforms static heuristics, offering a flexible, data-driven approach for dynamic production environments.

2.2.5.5 Proximal Policy Optimisation

Proximal Policy Optimisation (PPO) is a popular on-policy algorithm and a simplified successor to TRPO. It replaces the hard trust-region constraint with a clipped surrogate objective that penalizes deviations of the probability ratio $r(\theta) = \frac{\pi_\theta(a|s)}{\pi_{\theta_{old}}(a|s)}$ away from 1 beyond a threshold $\in$. The notation for the clipped surrogate loss function is:

$$L^{CLIP}(\theta) = \mathbb{E}\left[\min(r(\theta)A^{\pi_{\theta_{old}}}(s,a), clip(r(\theta), 1-\in, 1+\in)A^{\pi_{\theta_{old}}}(s,a)\right] \quad (2.9)$$

By avoiding second-order gradient steps and complex constraints, PPO is straightforward to implement, yet often delivers state-of-the-art performance on diverse benchmarks (Schulman et al., 2017). This balance of simplicity and stability has led PPO to become a default choice for many policy gradient applications, especially in continuous control and game domains.

PPO has been used widely in manufacturing research settings so far: Park et al. (2021) present an approach to solving the Job-Shop Scheduling Problem (JSSP) using Graph Neural Networks (GNNs) and PPO. The GNN is employed for representation learning, embedding the spatial structure of JSSP, while PPO trains an RL-based policy learning module to derive optimal scheduling actions. The proposed GNN-based scheduler outperforms traditional dispatching rules and RL-based approaches in various benchmark JSSP instances, exhibiting strong generalisation capabilities that allow it to transfer learned scheduling policies to unseen JSSPs without additional training. Rummukainen and Nurminen (2019) explore the application of PPO to the Lot Scheduling Problem, investigating scheduling manufacturing decisions on a single machine under stochastic demand, aiming to optimise costs and efficiency. A comparison with existing base-stock policies and Reinforcement Learning methods reveals that the PPO-based solution reduces the average cost rate by 2% over the best-known benchmark. Schneckenreither et al. (2021) investigate explores the use of PPO for dynamic short-term capacity planning in a make-to-order manufacturing setting. Unlike static methods, PPO adapts to disruptions like machine failures and fluctuating demand by optimising overtime allocation, balancing costs, and maintaining service levels. The study finds that PPO significantly reduces total costs compared to traditional fixed or random strategies, making it a more efficient and flexible solution for real-time production planning.

2.2.5.6 Deep Deterministic Policy Gradient

DDPG extends deterministic policy gradient methods (Silver et al., 2014) to deep networks, aiming at continuous action control scenarios such as robotics or autonomous driving. It trains two networks:

1. An actor $\mu_\theta(s)$ that outputs deterministic actions $a \in R^n$.
2. A critic $Q_\phi(s, a)$ that estimates the Q-value of (s, a).

DDPG is off-policy, using a replay buffer to sample uncorrelated transitions, and employs target networks to stabilise learning (Fujimoto et al., 2018; Lillicrap et al., 2015). Conceptually, it combines ideas from Q-learning (value function approximation) with policy gradient (direct adjustment of actor parameters) to handle large or continuous action spaces. While it can learn robust control policies, DDPG can exhibit sensitivity to hyperparameters and overestimation biases, inspiring refinements like TD3 (Fujimoto et al., 2018).

Being primarily suitable for deterministic continuous-time control processes, DDPG has been used for designing control policies for chemical mechanical polishing to reduce drift and shift issues due to material removal (Yu and Guo, 2020), urban traffic light control (Casas, 2017), capacity planning in supply chains (Tseng et al., 2025) and lighting control for a factory building to reduce the required power for lighting based on natural light illuminance (Kim et al., 2022).

2.2.5.7 Twin-Delayed DDPG

Twin Delayed DDPG (TD3) (Fujimoto et al., 2018) improves upon DDPG by tackling the overestimation of Q-values that can arise in approximate value-based RL. TD3's core innovations include:

1. Twin Critics: Two separate critic networks $\{Q_{\phi_1}, Q_{\phi_2}\}$ are trained. The target Q-value is computed by taking the minimum of both critics' estimates, mitigating positive bias in value estimates.
2. Delayed Policy Updates: The actor (policy) is updated less frequently than the critics to ensure the Q-functions have stabilised, reducing policy oscillations.
3. Target Smoothing: A small random noise is added to the target actions to reduce exploitable overestimation further.

By countering overestimation bias and smoothing updates, TD3 often outperforms vanilla DDPG, making it well-suited for continuous control tasks in industrial robotics (e.g., welding parameter tuning), chemical process control, or other real-world applications that require robust and stable policy learning.

Malus et al. (2020) address the problem of real-time order dispatching for a fleet of autonomous mobile robots by formulating the task as a multi-agent Reinforcement Learning problem. Using the TD3 algorithm, each autonomous mobile robot agent learns a bidding policy for incoming transport orders based on local observations, resulting in higher scheduling efficiency compared to conventional rule-based dispatch methods in simulation experiments. Within the topic of sustainable production, Qi et al. (2025) propose TD3 to optimise remanufacturing operations across heterogeneous multi-factory systems. It introduces a mathematical model that integrates U-shaped disassembly lines, diverse disassembly technologies, and Petri nets for task sequencing, with the goal of maximising profit while handling complex product disassembly and allocation. Experimental results demonstrate that the TD3 algorithm outperforms other Reinforcement Learning methods (like DDPG, SAC, A2C) and traditional solvers (e.g., CPLEX), especially in scalability and efficiency.

2.2.5.8 Soft Actor-Critic

Soft Actor-Critic (SAC) is an advanced off-policy actor-critic framework that introduces an entropy maximisation principle to encourage more robust exploration (Haarnoja et al., 2018). SAC optimises a stochastic policy $\pi(as)$ to maximise both expected return and an entropy term:

$$J(\pi)| = |\sum_{t=0}^{\infty} \mathbb{E}[r(s_t, a_t) + \alpha\mathcal{H}(\pi(\cdot|s_t))] \tag{2.10}$$

where $\mathcal{H}$ is the differential entropy of the policy, and α is a temperature parameter. By pursuing policies of higher entropy, SAC fosters broad exploration and avoids premature convergence to deterministic actions, a problem sometimes observed in DDPG-like methods. Empirical investigations by Haarnoja et al. (2018) show SAC outperforming previous continuous-control RL algorithms on benchmarks such as the MuJoCo suite (Duan et al., 2016). The algorithm's stability and off-policy data efficiency have positioned it as a strong baseline for continuous action domains, in research and practical robotics.

Since much like DDPG, SAC is advantageous for problems with continuous action and state spaces, its scholarly application has been primarily with process control problems: Coraci et al. (2021) implemented a SAC agent to control heating in an office building's Heating-Ventilation-Air-Conditioning (HVAC) system. Liu et al. (2024) proposed a robot navigation method using a SAC-based policy for autonomous forklifts/AGVs with LiDAR sensing. In the realm of continuous control assembly problems, SAC has demonstrated its effectiveness on robotic insertion tasks (Beltran-Hernandez et al., 2020; Zhao et al., 2020).

2.2.5.9 A3C and A2C (Advantage Actor-Critic)

A3C (Asynchronous Advantage Actor-Critic) introduced by Mnih et al. (2016), accelerates learning by running multiple parallel agents (threads), each interacting with its environment instance. Every thread maintains a local copy of the global policy π_θ and value function $V_\theta(s)$. After a short rollout, it computes gradients with respect to its local experience and updates the global (shared) model asynchronously. This asynchronous approach reduces the need for an experience replay buffer and helps decorrelate training samples, stabilising training in deep RL:

1. Each thread is an actor (runs the current policy to generate trajectories).

2. Each thread's local critic estimates state-value $V(s)$ or advantage $A(s, a) = Q(s, a) - V(s)$.
3. Gradients are summed asynchronously into the global model to update weights.

A3C achieved impressive results in both Atari and continuous control tasks, significantly accelerating training compared to single-threaded actor-critic approaches. Its parallelism can be especially beneficial in industrial scheduling (e.g., semiconductor dispatching), multi-AGV routing, and robust robotic assembly, where multiple simulated environments or scenarios can be run simultaneously to explore diverse states.

A2C (the "synchronous" version of A3C) takes essentially the same advantage-actor-critic approach but aggregates experiences synchronously from multiple parallel environment instances, then performs a single update step. This avoids the complexity of truly asynchronous gradient pushes and can be easier to implement in modern deep learning frameworks. While A2C and A3C often achieve comparable performance, A2C's synchronised updates can simplify debugging and reduce gradient noise. Both approaches leverage an advantage function to reduce gradient variance and typically do not require an experience replay buffer (unlike off-policy methods). Research studies have applied A2C to tasks such as industrial thermal process control, warehouse palletising, or real-time scheduling, demonstrating stable convergence and robust policy performance under dynamic or uncertain conditions.

In industrial contexts, Kim et al. (2020) use A3C to reduce unnecessary rearrangements in a ship block stockyard by dynamically deciding block placement and performing pre-emptive rearrangements. Barat et al. (2019) integrate an A2C agent with a simulated supply chain environment to optimise replenishment decisions, improving inventory management while minimising shortages and waste.

Publication I: Bibliometric Study on the Use of Machine Learning as Resolution Technique for Facility Layout Problems

3

3.1 Abstract

Facility Layout Problems (FLP) are concerned with finding efficient factory layouts. Numerous resolution approaches are known in the literature for layout optimisation. Among those, intelligent approaches are less researched than solutions from exact or approximating approaches. The recent surge of research interest in Artificial Intelligence, and specifically Machine Learning (ML) techniques, presages an increase in the usage of such techniques in FLP. However, previous reviews on FLP research induce that, to date, this trend has not yet emerged. Utilising a systematic literature review coupled with a k-Means based clustering algorithm, we analysed 25 relevant publication full-texts from an original sample of 1,425 papers. Our findings corroborate the statement that ML techniques have attracted substantially less research interest than most other resolution approaches. While a few papers used Unsupervised Learning algorithms directly as a solution to the FLP, Supervised and Reinforcement Learning were found to be practically irrelevant. ML usage was significantly higher in FLP-adjacent planning tasks such as group technology. Drawing on experiences with other NP-hard combinatorial optimisation problems in manufacturing research, we conclude that Reinforcement Learning is most promising for bridging the evident gap between FLP and ML research. Our study further contributes to FLP research by extending established classification frameworks.

Burggräf, P.; Wagner, J.; and Heinbach, B. (2021) "Bibliometric Study on the Use of Machine Learning as Resolution Technique for Facility Layout Problems," in *IEEE Access*, vol. 9, pp. 22569–22586, https://doi.org/10.1109/ACCESS.2021.3054563

B. Heinbach, *Reinforcement Learning-Based Planning of Factory Layouts*, Findings from Production Management Research ,
https://doi.org/10.1007/978-3-658-51554-6_3

3.2 Section I: Introduction

The Facility Layout Problem (FLP) is an important research stream within production research. An FLP is defined as the search for the most efficient arrangement of departments, i.e. facilities, in a plant area subject to different constraints while attempting to satisfy one or more objectives (Heragu and Kakuturi, 1997). According to Maganha and Silva (2017), FLPs can be located on the tactical hierarchical layer in organisations, provoking the assumption that sound decisions on facility layouts constitute a significant contribution to the economic success of manufacturing companies. Backing this notion is a general agreement on cost-saving potentials in a range of 10–30% for material handling cost-based operating expenses (Tompkins et al., 2010). Furthermore, a robust layout has an immediate measurable impact on the operational performance, as measured by manufacturing lead time, throughput rate, and work in process (Benjaafar, 2002). Another important driver to market survival in today's manufacturing environment is flexibility (Raman et al., 2009), which has sparked a research stream within FLP on its own right called Flexible Manufacturing Systems. The authors of this study posit that FLPs are an important organisational topic because a) proper factory layouts are required for an efficient and effective production process, and b) the frequency of FLP encounters is expected to increase with more customised manufacturing.

In the past years, we have experienced a surge of Artificial Intelligence (AI) reports in which AI algorithms have proven to outperform humans in complex, yet very specific tasks such as Go (Silver, Schrittwieser et al., 2017), complex online multiplayer games (Vinyals et al., 2019), Poker (Brown and Sandholm, 2018), hide-and-seek simulations (Baker et al., 2019) or emulations of old Atari games (Baker et al., 2019). Beyond such primarily demonstrative and trailblazing purposes, the question remains how the capabilities of AI can be used in value-adding industrial settings. It appears that widespread recognition of AI in practical real-life applications, especially in production or manufacturing research, is still pending. In Burggräf et al. (2018), we found that AI has predominantly been used in production for scheduling and quality control; however, we assert that AI will play a more vital role in higher-level production management planning and decision-making problems that lead to *cyber production management systems*. Several previous extensive reviews in the field of FLP have examined this research field concerning the *resolution techniques* used (Drira et al., 2007; Hosseini-Nasab et al., 2018; Leo Kumar, 2017; Maganha and Silva, 2017; Renzi et al., 2014). FLPs are NP-hard combinatorial problems (Kusiak and Heragu,

1987) which can be solved via approximating bio-inspired meta-heuristic algorithms yielding a fair solution rather than an optimal one, e.g. Genetic Algorithms, Particle Swarm or Ant Colony Optimisation, or recently Coral Reef Optimisation (Garcia-Hernandez et al., 2019). The adoption of Machine Learning (ML)—being a subset of AI—for complex real-life problems by both researchers and practitioners has made this field a dynamic area of research (Alzubi et al., 2018). Also, ML is known for its ability to handle many problems of an NP-hard nature (Monostori et al., 1998). Thus, it appears counterintuitive that the aforementioned reviews on FLP merely indicate a scarce use of intelligent approaches in FLP, while these are widely recognised as one class of these techniques. Models and algorithms that commonly star in Machine Learning terminology have been listed as one relevant tool, yet we do see a lack of a distinct drill-down in this specific domain. This paper seeks to tackle this limitation. Employing an extensive systematic literature analysis, the objective of this paper is to investigate to which degree AI topics, and more specifically different ML techniques, have pervaded the field of FLP. This study makes three main contributions to the extant literature.

1. We provide an in-depth examination of ML usage as a resolution method for Facility Layout Problems that goes beyond the available reviews in academic literature until thus far.
2. We extend previously known taxonomies of FLP resolution approaches by different meta-heuristics and Machine Learning algorithms.
3. The study highlights unexpected white spaces in ML usage FLP research.

The remainder of this paper is structured as follows: In Section 2, we will build up the AI framework that we use for our review. Section 3 will provide a detailed description of our chosen review methodology. The descriptive results of the literature analysis are presented and discussed in detail in Section 4. Results will be discussed in Section 5. Finally, Section 6 will provide our concluding remarks.

3.3 Section II: Background

3.3.1 Intelligent Approaches in Facility Layout Problems

Across the previous contributions in FLP research, there is no definite agreement on the definition of intelligent techniques. A large number of publications, especially of type review, seem to go back to the seminal review by Drira et al. (2007). While his framework includes *intelligence approaches* (cf. Fig. 3.1a), Drira et al.do not specify a second level here. Interestingly, Artificial Neural Networks are introduced as a class of problem formulation techniques rather than resolution approaches. The review itself does not have an own section on these approaches, nonetheless, Expert Systems (ES) are listed as relevant tools.

The framework used by Moslemipour et al. (2012) is notably similar to the previous one (cf. Fig. 3.1b) with the large exception that meta-heuristics are considered intelligent approaches here. Moslemipour *et al.* Moslemipour et al. further list Expert Systems, Fuzzy Systems, and Artificial Neural Networks (ANN) as tools.

Yet another picture is painted by Renzi et al. (2014). Here, AI is grouped under non-exact, i.e. approximating, approaches along with meta-heuristics (see Fig. 3.2). This taxonomy retains Fuzzy systems and ANN but drops ES. Renzi et al. identify four papers related to the use of ANN in reconfigurable manufacturing systems and 42 for cellular manufacturing systems. However, according to their analysis, these papers deal with cell formation and scheduling problems. No papers were identified as being relevant for cell layout problems.

Leo Kumar (2017) classified a total of five papers as being AI-relevant. However, he includes ANN in evolutionary techniques along with Genetic Algorithms (GA), Ant Colony Optimisation (ACO), and other algorithms more commonly denoted as meta-heuristics, which in turn, Kumar all groups as Artificial Intelligence.

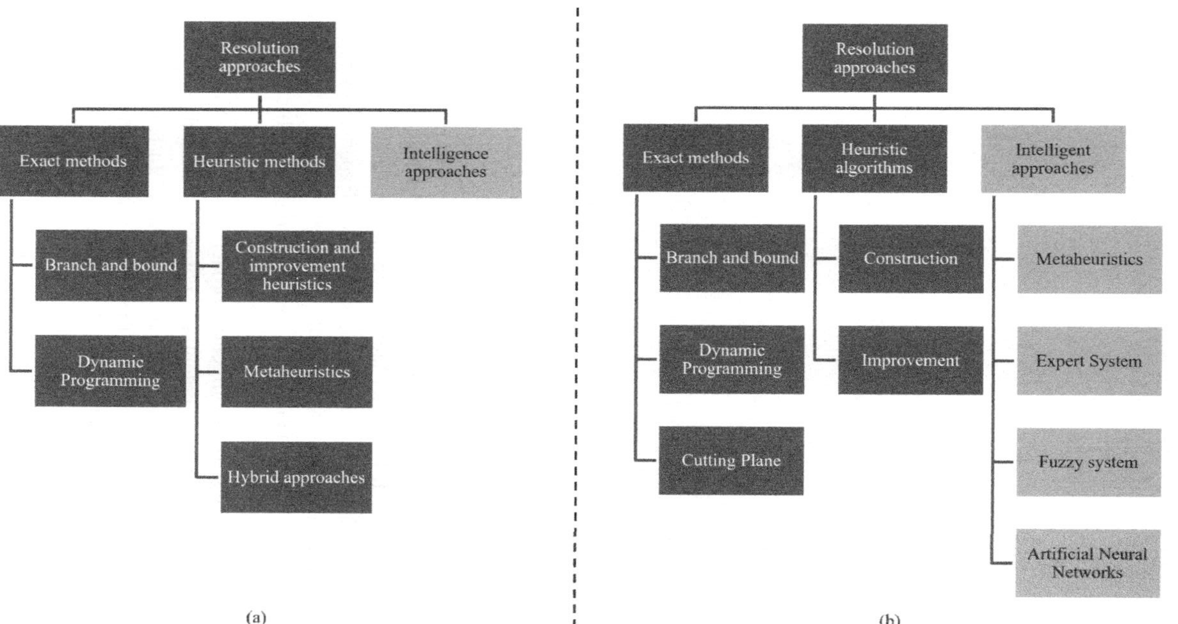

Fig. 3.1 Juxtaposition of the FLP frameworks used by Drira et al. (2007) (1a) and Moslemipour et al. (2012) (1b) to classify resolution approaches

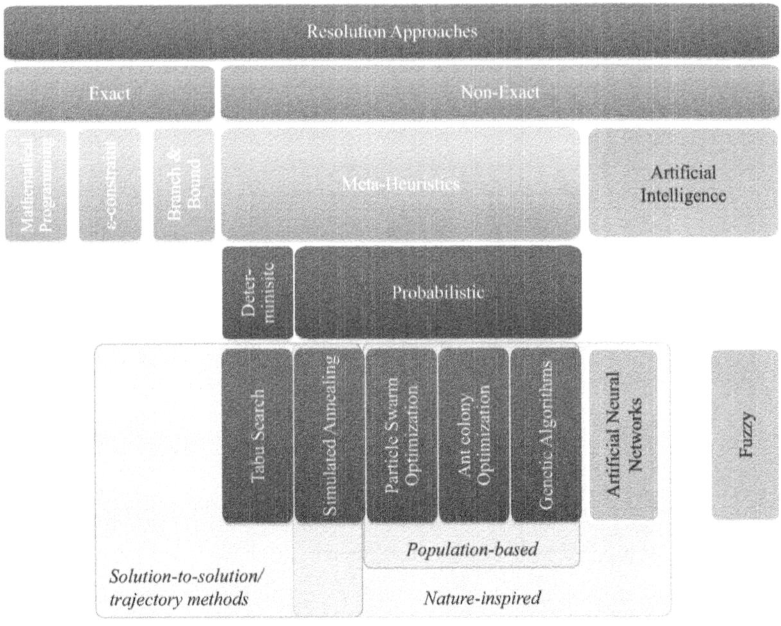

Fig. 3.2 Taxonomy of CMS techniques according to Renzi et al. (2014)

The probably most comprehensive framework can be found in Hosseini-Nasab et al. (2018) (see Fig. 3.3) where intelligent approaches are defined as a category of resolution techniques which is further subdivided into ES and ANN. In their extensive review, they find a total of five papers of which Expert Systems make up three and ANN the remaining two. Not only is the low number surprising by itself, but the authors also merely name these two concepts as a sub-class and do not provide further details as to their usage.

The richness of taxonomies for intelligent approaches certainly leads to wide interpretability of results and thus to potentially competing insights. We do not intend to draft an additional and entirely new framework. Instead, due to its recency and comprehensiveness, we extend the recent classification scheme in Hosseini-Nasab et al. (2018) for intelligent approaches such that it becomes workable for our review on ML techniques in Facility Layout Problems.

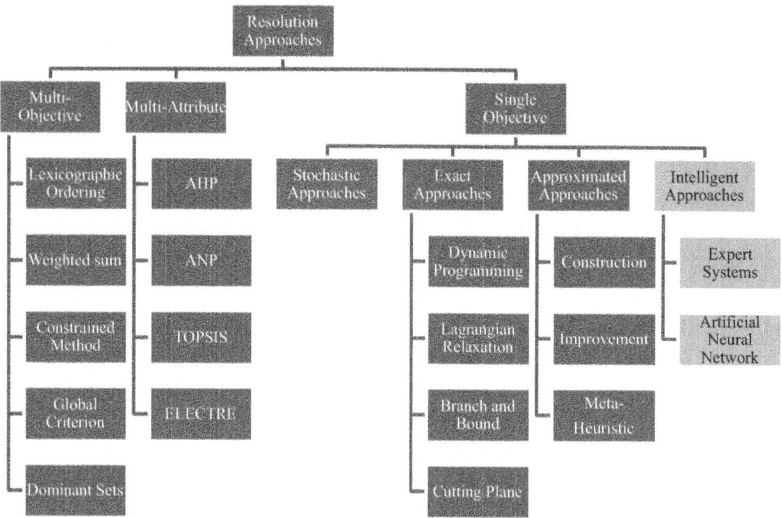

Fig. 3.3 Resolution approach classification used by Hosseini-Nasab et al. (2018)

3.3.2 A Conceptual Framework for Coding Intelligent Approaches

ML techniques belong to a problem representation framework called *sub-symbolic AI* (Gentsch, 2019). Following a bottom-up, i.e. data-driven, approach, sub-symbolic AI attempts to create structures that can learn intelligent behaviour through information-processing mechanisms (Turing, 1948). The counterpart to sub-symbolic AI is *symbolic AI*, which is originally defined as a system that has sufficient means for general intelligence by symbol manipulation (Newell, 1980). Here, knowledge is encoded in terms of explicit symbolic structures, and inferences are based on handcrafted rules that sequentially manipulate these structures (Miikkulainen, 1994). Preparatory work on our review study, most notably the scoping review as described in section III-A, showed that the umbrella term "Artificial Intelligence" is inherently chosen over usage of the term "Machine Learning". We, therefore, pick the starting point of AI as the top-level node in our taxonomy, followed by the two branches symbolic and sub-symbolic AI. As indicated above, the three concepts that have attracted the most research interest within intelligent resolution approaches regarding their occurrence in conceptual review frameworks are Fuzzy systems, Expert Systems (ES), and Artificial

Neural Networks (ANN). ES are a well-studied member of symbolic AI (Miikkulainen, 1994). However, research interest in ES has been declining continuously in operations research (Burggräf et al., 2018; Kobbacy et al., 2007; Kobbacy and Vadera, 2011). Considering ML definitions by Bellman (1957), Samuel (1959) and Gentsch (2019), a key trait in ML, is the capability to *independently* learn and be able to find solutions to previously unseen situations. As ES are a one-to-one representation of expert domain knowledge which are known to lack adaptation capabilities in new situations, we exclude ES from our synthesis scope. Nonetheless, as previous reviews have indicated, we acknowledge that ES were an important research topic in the past and therefore retain the concept for coding purposes. Regarding Fuzzy systems, we argue in line with Drira et al. (2007) and Hosseini-Nasab et al. (2018) that Fuzzy numbers and Fuzzy systems ought to be classified as data types for layout formulations. This means that any variable used in the FLP can either be notated as crisp, as stochastic, i.e. sampled from a distribution, or as a degree of membership, ranging from 0 to 1, in a fuzzy set. Thus, as opposed to some authors before us, we will not include Fuzzy systems in the class of intelligent resolution approaches.

Lastly, ANN have a long history in AI since being introduced in 1943 by McCulloch and Pitts (1943). While they have recently gained traction thanks to breakthroughs in some fields like image recognition (Hardy, 2016; Le, 2013), they do not comprise the only available model in ML. In fact, a substantial number of different models and algorithms have been designed. ML can be subdivided into the classes Supervised Learning (SL), Unsupervised Learning (UL), and Reinforcement Learning (RL) (François-Lavet et al., 2018; Wuest et al., 2016). *Supervised Learning* is a data-driven learning paradigm in which inputs are mapped to outputs. Methods include, among others, Artificial Neural Networks, Logistic Regression (LR), K-Nearest-Neighbour (k-NN), Support Vector Machines (SVM), Random Forest (RF), Decision Trees, and Bagged Trees (Caruana and Niculescu-Mizil, 2006). *Unsupervised Learning* on the other hand is a branch of Machine Learning method that aims at finding hidden patterns in unlabelled data instead of mapping known inputs to known outputs (Hinton et al., 1999). Typical examples of UL are clustering (e.g. via k-Means), association rules, and self-organising maps (a specific architecture of ANN) (Sammut and Webb, 2011). *Reinforcement Learning* mimics human decision-making by having an agent gradually collect information from interaction with its environment. For each action performed in a distinct state, the agent collects a reward, which, over the course of several iterations, shapes a policy that describes the optimal action to take for each state (Sutton and Barto, 2018). Prominent RL algorithms are Q-Learning, state-action-reward-state-action (SARSA) (Sutton and Barto, 2018) or Deep-Q-Networks (DQN) (Mnih et al., 2015), to name but a few. In addition to those three, *Deep Learning* (DL), uses multiple layers to progressively extract higher-level features from raw input data (Deng, 2014), typically using Neural Network architectures with a significant amount of hidden layers. DL can be used both in a supervised, semi-supervised or unsupervised fashion (LeCun et al., 2015). Plus, Google's DeepMind has fostered attention *in Deep Reinforcement Learning* (DRL) by outperforming human players in ATARI games (Mnih et al., 2015). Showing links to all other branches, we consider DL as a cross-sectional collection of methods. However, for ease of visualisation, we grant it a separate branch in our framework and highlight the aforementioned particularity with an asterisk (*).

Artificial Neural Networks have been proposed in a multitude of architectures, a sample of which is presented below as listed by Leijnen and van Veen (2019). The assignments to learning paradigms in square brackets were added by the authors of this paper.

- Feed Forward Neural Networks (FF) and multilayer perceptrons (MLP) [SL]
- Recurrent Neural Networks (RNN) [DL]
- Long Short-Term Memory (LSTM) [DL]
- Autoencoders (AE) [UL]
- Hopfield Networks and Boltzmann Machines (HN) [UL]
- Convolutional Neural Networks (CNN) [DL]
- Generative Adversarial Networks (GAN) [UL]
- Kohonen Networks, or self-organising maps (SOM) [UL]

A synthesis of the above discussion is depicted in Fig. 3.4. This figure shows the conceptual framework for ML in FLP that we use to code retrieved articles. For our study, we adjust the definitional review scope for intelligent approaches by enriching the previously used concept of ANN, by dropping Expert Systems due to their lack of recent research focus, and by incorporating our understanding of ML approaches in the form of lowest-level tree nodes. Drawing from Renzi et al. (2014), we further added common meta-heuristic methods as concepts for a better overview.

Note that the visualisation in Fig. 3.4 is not meant to be exhaustive but seeks to underline the complex structure and the diverse nature of currently available and common ML concepts. Hence, only those concepts discussed above are shown in the figure. Since we made no changes to the framework in Hosseini-Nasab et al. (2018) regarding exact and stochastic approaches, we exclude these branches from our visualisation for better legibility. Our additions to approximating and

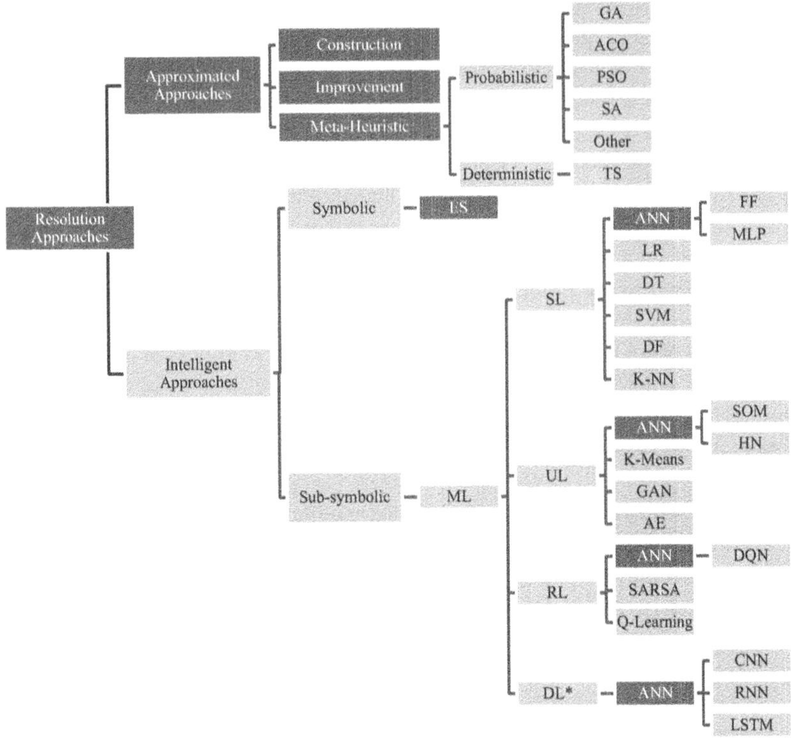

Fig. 3.4 Synthesis of the ML framework used for coding in this review: Additions to the previous framework are shown in green

intelligent approaches are highlighted in green colour. The following section will present in detail how the framework is leveraged in our review.

3.4 Section III: Methodology

A systematic literature review is a research methodology to "review [...] research literature using systematic and explicit, accountable methods" (Gough et al., 2012, p. 2), that has become the gold standard to synthesising findings from several studies investigating a similar question regardless of the discipline where it is used (Boland et al., 2017). Systematic reviews tend to have a narrow research question, which, combined with strict quality criteria, result in a reasonable number of studies to be included in the review whose results add up to answering the research question. This sort of review can be termed aggregative (Gough et al., 2012) and qualifies as conceptual rather than empirical research.

To provide a comprehensive overview of the state-of-the-art of research on Artificial Intelligence, and specifically Machine Learning algorithms, in kijnm-factory planning, this paper adopts the methodology for conducting systematic literature reviews proposed by Boland et al. (2017). As it became apparent early on that the review would face a substantial number of publications, we extended this methodology by an algorithmic clustering step, see Fig. 3.5. The methodological details of each step are presented in the following sub-sections.

Section	Step	Content
Scoping (3.1)	1	Perform scoping search (refine research question, set inclusion criteria)
Lit. Search (3.2)	2	Search literature (includes removal of duplicates)
Clustering (3.3)	3	Perform cluster analysis
Data Analysis (3.4)	4	Screen titles and abstracts ("Stage 1 Screening"), apply inclusion criteria
	5	Obtain papers (full-texts)
	6	Select full-texts ("Stage 2 Screening"), apply inclusion criteria
	7	Quality assessment
Results and Discussion (4 and 5)	8	Data extraction
	9	Analysis and synthesis
	10	Write-up

Orginal methodology by Boland et al. (2017) Authors' extended methodology

Fig. 3.5 Extended SLR methodology building up on Boland et al. (2017)

Furthermore, following the advice Gough et al. (2012), we assembled a review team consisting of this paper's authors to increase inter-rater reliability for the data analysis phases.

3.4.1 Scoping Search

The scoping review we conducted indicated a large number of returns in the databases, ranging beyond a six-digit figure. This large number of records exceeded the research team's capacity to screen and retrieve. As proposed by

Gough et al. (2012), the scoping search motivated the authors to employ automatic clustering as a text mining approach. Therefore, we included Step 3 (Cluster Analysis) in the review process. The purpose of this step is to partially automate the classification process in Stage 1 Screening (Step 4).

3.4.1.1 Research Question

As outlined above, the purpose of this paper is to investigate the degree to which ML techniques have been used to solve Facility Layout Problems. Thus, we formulate our research question as follows:

RQ1: *How have different Machine Learning algorithms been used as resolution techniques for Facility Layout Problems?*

3.4.1.2 Inclusion Criteria

To identify a body of literature that is relevant to answering the research question, we established the following quality criteria:

1. Include peer-reviewed academic work, such as books, journal articles, and conference contributions
2. Include theses and dissertations
3. Omit other grey literature (e.g. magazines, commercial websites, company white papers, patents)
4. Exclude publications not pertinent to Facility Layout Problems
5. Exclude all non-English publications
6. Include only articles newer than 1987 (review scope of Hosseini-Nasab et al. (2018))

For the detailed analysis, we set the following inclusion criteria (IC) to be applied to all references short-listed for Stage 1 Screening:

- *IC1:* The paper is relevant to the research area "Facility Layout Problems"
- *IC2:* The paper indicates the use of any of the ML techniques from the framework in II-A

3.4.2 Literature Search

3.4.2.1 Search Terms

Figure 3.6 displays the combinatorics of terms commonly used in FLP research. The most common term is "Facility Layout Problem", as highlighted in blue. However, "plant layout design" (Chin-Sheng and Kengskool, 1990), "facility layout planning" (Potocnik et al., 2014), "site layout planning" (Hammad et al., 2016), "space layout" (Lobos and Donath, 2010) or simply "layout problem" (Martens, 2004) can be found as well. This combinatorics indicates a richness in research that needs to be addressed in the systematic search. There are two important considerations for the search strings. First, a search string including three components as depicted below proves to be difficult to enter or resolve in some of the databases used in this research. A uniform way to address this is a long string of OR-combinations, which may be inapplicable due to character input restrictions. However, any database query for the former two components should pick up publications including all terms in the third component. Using all three components in Fig. 3.6 also increases the risk of missing an important stream not involving any of the terms. Therefore, we chose to eliminate this part from the search string and only regard the combinatorics from the left-hand side of the dashed line. Second, the inclusion of the first component is vital to address the correct research communities. Excluding terms such as "facility", "factory", "plant" and the like showed to pick up publications from hospital and operating theatre designs or integrated circuit layout design. While there may be relevant knowledge on RL applications in those adjacent fields, our initial scoping revealed an extensive literature base already for the domain of facility layouts. Therefore, we opt to exclude publications not having a distinct manufacturing focus. Yet, we acknowledge that a thorough search in these fields whilst highlighting similarities and congruencies in the approaches or logic can be a worthwhile topic for future research.

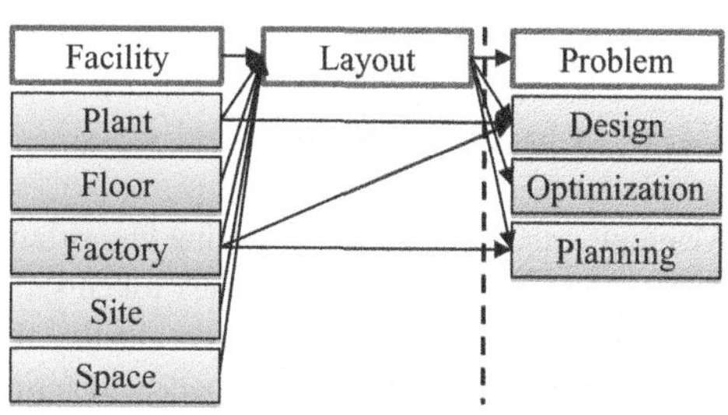

Fig. 3.6 Overview of relevant search terms

3.4.2.2 Databases

The following nine databases were queried with the search strings above: ScienceDirect, Web of Science, JSTOR, EBSCO Host, Emerald Insight, IEEE Xplore, ProQuest, and ACM Digital Library. This selection is based on our previous good experience with these databases (see (Koke and Moehler, 2019; Weißer et al., 2020)). Furthermore, we tackled the file-drawer bias by searching in OpenReview for potential AI-related contributions in FLP that may have failed a peer-review process.

3.4.2.3 Dataset Preprocessing

As the scoping indicated, our search strategy yielded a large number of retrievals. Therefore, in a first step, the full dataset has been cleaned from all irrelevant entries such as front or back matters, volume contents, errata, glossaries, forewords, keyword indices, and the like. Titles, abstracts, and keywords were changed to lower case to facilitate the identification and removal of duplicates. In the last step of preprocessing, we recreated a title-abstract-keyword search that was not supported by most databases. The purpose of this approach is to exclude entries which do include the search terms but without close relation to the search field. Using a short macro, we re-applied the above search string to the database fields "Title", "Abstract" and "Keywords", thus eliminating a large number of artefacts we consider false positive entries in our dataset. The result of this step is a cleaned dataset ready for Stage 1 Screening.

3.4.2.4 Snowball Sampling

Given that *snowball sampling* is a powerful data collection tool in systematic reviews which can contribute up to half of the relevant references in a study (Greenhalgh and Peacock, 2005), we applied this sampling technique to the review articles from 3.3.1. We performed snowball sampling during a longer period between September and October 2019. The references obtained via this way were added to the pre-processed dataset.

3.4.3 Cluster Analysis

We employ the NLP-based clustering algorithm designed by Weißer et al. (2020). This k-Means based algorithm uses tokenisation (word separation), stop words removal and TFIDF vectorisation to identify the most relevant words per cluster that describe its thematic relevance. Clustering was performed on the "Title" column of the cleaned dataset for Stage 1 Screening. From the previous reviews we synthesised 19 possible clusters consisting of different exact, meta-heuristic and intelligent techniques and one cluster for un-classifiable papers. Accordingly, we set the algorithm to creating 20 clusters. Thus, we selected a *kMax* for the Elbow method of $kMax = 20$, and an explained variance of 0.7. The algorithm was tasked to indicate the five most relevant top words in each cluster.

3.4.4 Data Analysis

3.4.4.1 Stage 1 Screening

For the Stage 1 Screening procedure, the remaining papers in the dataset were screened manually. To increase inter-rater reliability, the final dataset for this step was split evenly and at random and distributed among the three members of the research team. Papers were presented to the researchers only via their title, abstract, and keywords. Each researcher coded their respective papers according to the framework presented in Fig. 3.4 (i.e. categorical coding). The concepts within the "Exact" branch of Fig. 3.3 were united under that term. The multi-attribute concepts describe group decision techniques and were thus coded as "qualitative". Along with the lower-level entries in this conceptual framework, the labels "unknown" or "none" (for publications, e.g. of conceptual nature, which did not clearly report on any resolution technique), "not FLP" (for papers that discuss layout issues but not within the facility problem domain, i.e. not meeting IC1) as well as "review" (for publications of type review which are assumed to refer to the majority of publications included in this dataset). We determined that for references coded with an ML-relevant concept from the above framework both inclusion criteria IC1 and IC2 applied. The respective papers thus were carried forward to Stage 2 Screening.

Upon completion of the first coding round, a random subset of references at a size of 20% of each researcher's dataset (i.e. approximately 65 publications each) was re-distributed to another researcher in the team for a second blind coding. Any mismatches in the coding results were mediated by the third researcher. This number of interventions was considered sufficiently low to warrant a 20% sample size as sufficiently large.

By the end of the above screening process, we pooled the relevant coding results from the two sources clustering and manual screening.

3.4.4.2 Obtain Papers

All papers flagged relevant for Stage 2 Screening could be obtained as full text. Each paper was given a unique two-digit ID for further coding processes.

3.4.4.3 Stage 2 Screening

In Stage 2 Screening, initial coding results from Stage 1 Screening were verified. Full texts were screened by the lead researcher. In a separate slide deck, a specific slide was maintained for each publication. Full-text snippets were collected in support of or in opposition to both inclusion criteria and new codes were applied where necessary. As a result of an emergent low-precision pattern, two different final codes were applied to the publications in review: papers to which IC1 and IC2 applied were assigned the code "Prio 1". To enrich the final synthesis sample, papers which reported on a distinct ML technique from the framework, albeit not meeting IC1, were labelled "Prio 2". For these two classes, the publications were labelled according to the Machine Learning paradigm in question, i.e. Supervised Learning, Unsupervised Learning, or Reinforcement Learning. The final subset containing all "Prio 1" and "Prio 2" papers was used for the synthesis of our findings.

3.5 Section IV: Description of Analysis Results

This section reports the results of each review step from Section III according to the PRISMA method (Moher et al., 2009). The numbers reported hereafter can be traced using Fig. 3.7.

The literature searches were performed in late September 2019. We identified a total of 11,851 publications across all nine databases. Only OpenReview did not incur any results. 1,903 irrelevant entries were removed. Another 191 entries were removed for being duplicates. In data pre-processing, we excluded an additional 6,811 publications through the recreated title-abstract-keywords search. Furthermore, before entering the Stage 1 Screening process, we examined all titles, and where necessary the abstracts, too, as to the relevance to FLP (IC1). IC1 did not apply to 1,601 entries, leading to their removal. This resulted in a final dataset for screening consisting of 1,303 references.

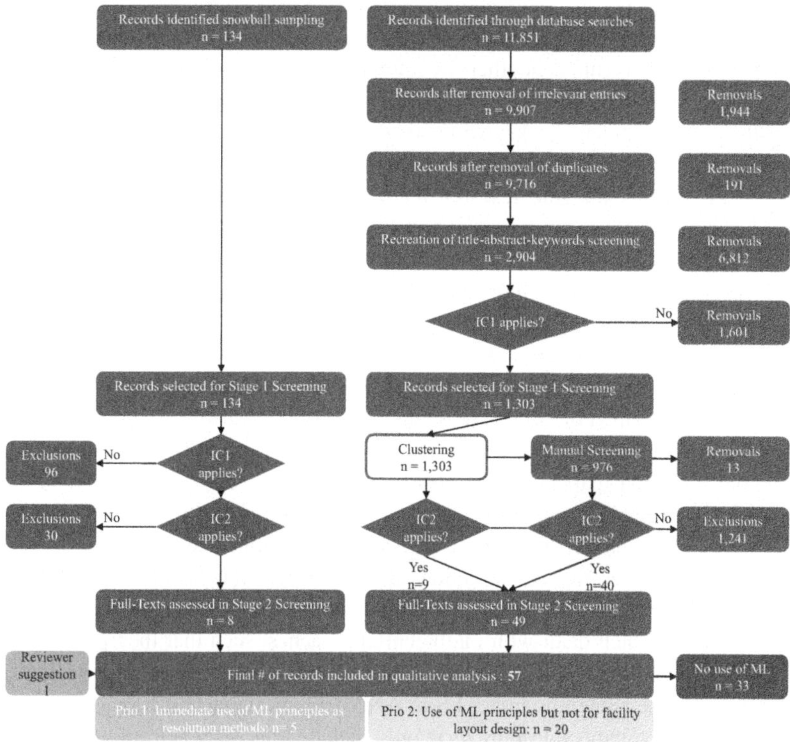

Fig. 3.7 Stepwise report for all removals during the review process

The dataset was first fed into a clustering algorithm which assigned 327 papers into clusters relevant for further analysis: The algorithm identified a cluster for the words "artificial", "intelligence", "neural", "network", "system", containing a total of 20 papers. Out of the 20 papers considered relevant by the clustering, nine progressed further to Stage 2. Three papers did have an ML focus but were excluded for being review articles. The remaining eight papers were re-coded according to the title and abstract information. The algorithm grouped five further clusters which could be unambiguously mapped to the meta-heuristic concepts known in FLPs: Genetic Algorithms, Ant-Colony Optimisation, Particle Swarm Optimisation, Tabu Search, and Simulated Annealing. A total of 307 papers were assigned to these clusters. These were screened again manually to verify the inclusion choices made by the clustering algorithm. The classification could be affirmed for 177 papers, the remainder was re-coded accordingly. As a result of the clustering, 976 papers remained for manual assessment by the research team.

In Stage 1 Screening, another 13 references were removed manually: eight were duplicates, two were notices of articles being retracted, and for the remaining three no abstracts or keywords could be retrieved. On top of the nine ML-relevant references picked up by the clustering, Stage 1 Screening yielded another 40 papers that the research team considered potentially relevant. The 1,241 papers excluded for not satisfying IC2 were coded with a label describing the resolution method used.

From the snowball sampling screening of the previous reviews, depicted on the left-hand side of Fig. 3.7, we obtained a total of 134 additional papers, 27 in the backward direction (cited articles) and 107 in the forward direction (citing articles). In the forward search sample, we identified three new papers that were potentially relevant. For backward sampling, we revisited the papers that were used in the previous analyses. The paper of Renzi et al. (2014) contained 17 potentially relevant papers. Although the original study concluded that none of their reviewed publications dealt with layout generation, we marked two papers for further examination in Stage 2 Screening. As outlined before, Leo Kumar (2017) used a diverging AI definition. Based on our framework, we found no ML-related papers in this sample. Hosseini-Nasab et al. (2018) reported five AI-related papers which dealt with intelligent approaches, such that they progressed to Stage 2. On closer scrutiny of their references 186 and 133, we cannot confidently assign the content to either ES or ANN and thus contest the above number of five AI-related papers, leading us to only include three papers from this sample to Stage 2 Screening. Thus, snowball sampling yielded a total of eight papers we considered relevant for full-text analysis.

In Stage 2 Screening, we assessed a total of 57 papers regarding IC1 and IC2. We found five papers that reported on ML techniques used as a resolution method for FLPs (Prio 1). Given that we considered this a relatively low number for a proper narrative analysis, we flagged another 19 papers which dealt with ML techniques yet were used primarily as support functions for layout problems (Prio 2). The papers not labelled with any of these two categories were re-coded according to the actual resolution technique used, if any. Ultimately, 24 papers were included in the synthesis in the discussion section.

Figure 3.8 shows the final coding results of our review study. Not shown in this figure are references coded as "reviews" and those excluded for not meeting IC1. The data show that the cumulative figure for intelligent approaches, marked as the green bar, is marginal compared to all other approaches. In line with the review study conducted by Hosseini-Nasab et al. (2018), the total number for approximating approaches is by far greater than that for exact ones, followed by stochastic and lastly intelligent approaches. The dominant role of genetic algorithms in solving FLPs is incontestable.

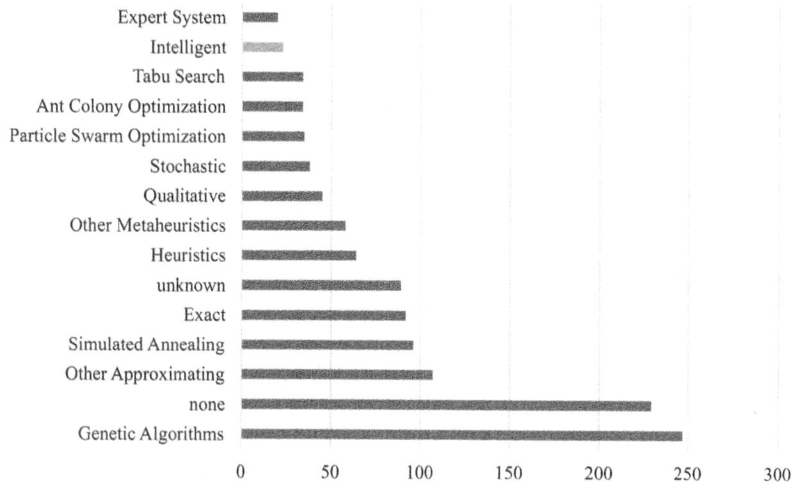

Fig. 3.8 Aggregated results for resolution methods after Stage 2 Screening (n = 1212)

As shown in our framework, we have purposefully excluded Expert Systems from our scope of analysis for not being relevant to Machine Learning. Our analysis shows that, albeit by a narrow margin, Machine Learning approaches have overtaken Expert Systems in terms of cumulative numbers around the year 2010 (see Fig. 3.9). ES stagnation is in line with the aforementioned decline in research interest. The data in Fig. 3.9 also show that intelligent approaches were outnumbered by qualitative, stochastic, and several approximating approaches in the early 2010 s as well.

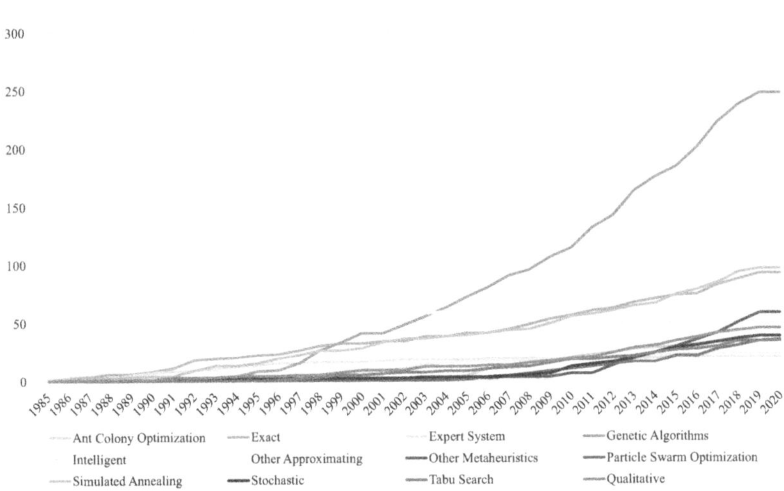

Fig. 3.9 Time series analysis showing cumulative results for resolution methods

3.6 Section V: Results

3.6.1 Study Findings

In the following, we will discuss our findings grouped by the learning paradigms from our research framework in Fig. 3.4. Deep Learning is treated as a cross-sectional paradigm and will thus be presented accordingly within the other three, where applicable.

3.6.1.1 Unsupervised Learning

Within Unsupervised Learning, we found three predominant concepts in FLP research: clustering techniques, self-organising maps (SOM), also referred to as Kohonen Networks and Hopfield Networks.

In Kobayashi et al. (2001) and Ueda et al. (2002), a potential fields modelling method is used to realise self-organisation in semiconductor manufacturing. Machines are represented by the nodes of the neural network and assigned machine properties such as "idle", "in process", "malfunction", "maintenance" and "buffer with product". Then, attraction and repulsion fields, the latter type to avoid machine overlap, are generated for the machines according to these properties. Attraction field attributes for transporters and buffers are modified by the requirements of the product. As a result of their simulation, the total travel distance for the product being produced has been minimised such that the most frequented machines in the manufacturing process were orbited by their respective support functions.

Similarly, Tsuchiya et al. (1996) employ an artificial two-dimensional maximum neural network to solve a QAP where N facilities are to be placed on an $N2$ location array. The number of nodes in the ANN matches the number of locations. The ANN provides a gradient descent method to minimise the fabricated energy function. All nodes' activation functions are step functions between 0 and 1, to be activated for the node that has the strongest weight at a given time step. Weight strength is determined via Manhattan distance, i.e. shorter distances between facilities, and thus lower transport cost, yield higher weights.

Zhang et al. (2002) propose the use of Hopfield networks for construction site layout planning. The function of the Hopfield network is to assign temporary establishments (such as tower cranes on construction sites) to pre-defined locations available for layouts, using the knowledge passed from an Expert System and info input by a user. The construction sites are mapped to the network nodes and the energy function consisting of distance metrics is being minimised via forward computing until convergence is achieved and an optimal layout thus found. However, the paper does not provide a clear description of how the above-mentioned assignment is performed in detail. It appears that the framework is run once; algorithmic support, such as Simulated Annealing, to find a stable state through adjustment of network weights is proposed yet not explained further.

Such algorithmic support is described by Yeh (2006): using a hospital case study with 28 facilities formulated as quadratic assignment problem (seven sites on four floors), the authors extend previous approaches to an Annealed Neural Network which combines characteristics of the Simulated Annealing algorithm and the Hopfield Neural Network. The purpose is to maintain the rapid convergence capabilities of the Neural Network while preserving the solution quality afforded by Simulated Annealing. This way, the Hopfield Network's major disadvantage of not being able to find the global energy minimum in a single run, but settling with the nearest local minimum instead, is overcome. The Hopfield Network recomputes the permutation matrix, which describes the strongest links of assignment of facilities to sites. Such an assignment is determined by a local energy minimum, calculated as a function of layout preference (space requirements), interactive preference (adjacency), and constraint violation (penalty). According to the authors, this approach makes it independent of the size of the search space, unlike many other approaches.

The study of Vakharia and Wemmerlöv (1995) analyses the impact of dissimilarity measures and clustering techniques on the quality of solutions in the context of cell formation. Instead of SOM, they propose a clustering algorithm called "set merging" where the smallest entity considered during the merging process is the cluster rather than the objects within each cluster. As with the examples mentioned before, layout resolution is not within the scope of their contribution.

In contrast to the examples provided above, where self-organising maps (SOM) can be regarded as a resolution technique by creating layouts, this type of algorithm has found wide application as a support function in the above-mentioned group technology. This type of application is within the scope of Prio 2 papers. Examples of this can be found in (Ateme-Nguema and Dao, 2009; Berlec et al., 2014; Jang and Rhee, 1997; Potocnik et al., 2014; Tateyama et al., 2003a, 2003b). For the same purpose, Tateyama and Kawata (2004) provide an extension to their previous SOM by using an Adaptive Resonance Theory-based clustering, which yielded better grouping results. Another approach that reports the use of SOM is presented by Sahragard and Bashiri (2016). Here, SOM is used in conjunction with ALDEP, a member of the class of construction algorithms. While they employ ALDEP for the actual layout resolution (and the paper has been coded by the researchers as such), SOM is used to tackle problem size issues as described in the introduction section. As the ALDEP algorithm can only handle problems with up to 14 facilities, SOM groups facilities into clusters, and each cluster's individual layout is created via ALDEP.

3.6.1.2 Supervised Learning

The most remarkable and unexpected finding regarding Supervised Learning approaches is that we found none which satisfies both our IC1 and IC2. For those satisfying IC2, we found evidence for supervisors to act either as classifiers (e.g. an expert rating surrogate (García-Hernández et al., 2014)) or as regressors (e.g. in the case of hoisting times (Tam and Tong, 2003)) without a distinct tendency towards one of these two problem classes. For those that do not satisfy IC1, contributions either dealt with cell formation problems in cellular manufacturing research, quite similar to our analysis on Unsupervised Learning approaches, or relied on other algorithms as resolution techniques.

A prominent approach in ML research for FLPs is to combine Neural Network capabilities with meta-heuristic approaches. We found that for these approaches, the key functionality of the ANN used is the capability to represent non-linear relationships to compute input parameters for the construction or optimisation algorithms in use. Tam and Tong (2003) and Azimi and Soofi (2017) both present solutions that make use of Genetic Algorithms and Neural Networks. In Tam and Tong (2003), a simple back-propagation network (five input nodes, one fully connected hidden layer, one output node) predicts hoisting times (supply and return) of tower cranes on construction sites. The network is trained with 1,000 cases from six projects, whereas the test datasets consisted of other data points from a further project. The GA in the setup creates a layout based on a population and evaluates their fitness by computing total travelling times using the predicted hoisting times.

Azimi and Soofi (2017) make use of ANNs to estimate the make-span of production jobs in different layout scenarios: a number of different layouts are generated and the make-span is determined using discrete-event simulation. These input-output pairs, with the input being a vector of machine locations and transporter allocations, are used to train an ANN to approximate the objective function. However, a seemingly low number of 24 layout simulations was used for training. As in Tam and Tong (2003), the layouts are designed by a GA, NSGA-II in this case. The authors claim that the ANN serves as chromosomes fitness function evaluator, but details of how the mechanism to evaluate the material handling cost-based fitness function works are not provided.

A GA is employed to generate layouts in an optimisation phase in García-Hernández et al. (2014) as well. The resulting set of layout solutions is then presented to an expert for evaluation according to the factors material flow, adjacency requirements, and aspect ratios using a five-point Likert Scale. The evaluation results are then used to train an ANN, which will act as a surrogate for the expert, primarily to avoid fatigue. The training and validation datasets used here seem sufficiently large (365 and 181 cases for two tests at an approximate 50% split for training and test data). When subjected to test data or new layout instances, the trained model will predict the mark that the expert would have given to that layout by applying a softmax function on the five output classes (corresponding to the 5-point Likert evaluation) at an accuracy ranging between approx. 80% and approx. 91%, depending on which expert evaluates the ANN predictions. Thus, the key to this contribution is to incorporate human knowledge into the FLP evaluation phase.

The usage of expert ratings in combination with Machine Learning can be found in Tayal et al. (2020) as well. Here, a pool of layouts is generated using SA, Clonal Selection Algorithm, and Firefly Algorithm. Relevant criteria are identified by submitting them to a rating by company experts. Then, an Unsupervised Learning approach leveraging Principal Component Analysis (PCA) is used to obtain a subset of uncorrelated criteria. Following an efficiency analysis using Data Envelopment Analysis, the authors use Supervised Learning based on training linear and logistic regression models to predict layout rankings. Model optimisation is performed with k-fold cross-validation.

Further, Deep Learning systems that include expert knowledge are also presented by Chung (1999a) and Chung (1999b). Here, neuro-based Expert Systems using Bidirectional Associative Memory (BAM) Neural Networks are proposed. BAM Neural Networks are a form of Recurrent Neural Networks, a member of the Deep Learning class, that have a simple architecture, consisting of two fully connected layers, and need fewer data points than back-propagation. Each neuron in a layer stands for one abstract concept (assertion) of a layout rule. The key is to translate verbal if-then rules from layout requirements into a matrix reformulation of weights (i.e. the memory) and match the rules to layouts. A variety of training examples can be generated from empirical simulations, historical data, or layout software programs. The BAM system analyses the layout clues, including closeness relationship, relative position, area size, and site constraints, to identify patterns and relationships that may subsequently lead to rules for laying out facilities. While this system can aid a planner by classifying a layout according to specified input requirements, i.e. estimate the layout's fitness, this approach does not in itself create layout alternatives other than a generalisation from training examples. The layout design is accomplished by construction algorithms such as CORELAP.

Group technology problems without immediate layout considerations have also played a role in Supervised Learning. Chen and Sagi (1995) designed a Neural Network that predicts certain manufacturing cell design parameters such as layout strategy, number of operators needed, part mix, and production sequence. The Neural Network design consists of two cascaded networks: a design Neural Network takes performance measures as input and outputs cell configurations. These, in turn, are fed into a control Neural Network, which will output cell control function ratings. Training occurs with pairs of performance measures known a priori and cell configurations, as well as complexity requirements for control functions, respectively. Layout strategy is mentioned as one predictor, yet this does not refer to layout types (i.e. transfer line, loop, warehouse, etc.) but to a binary classification of machines grouped vs. using two separate lines for the examined products to be manufactured. Layout design is not within the scope of the paper.

Rao and Gu (1995) used a combination of Unsupervised and Supervised Learning to solve a cell configuration problem. First, a cluster centre-seeking algorithm similar to k-Means takes a production flow analysis chart as input and identifies cluster centres which are most disparate from each other according to the target number of clusters specified by the user. The cluster centres are input to a Neural Network. While the authors state that the network uses "reinforced learning", as "vigilance values are modified during the training process to inflict punishment (negative feedback) for an erroneous classification" (p. 1058), the description of top-down and bottom-up weight adaptations implies that this refers to back-propagation. Moreover, it is stated that the network then acts as a classifier that maps new input patterns to the existing exemplars. While these reflections allow us to classify this contribution as Supervised Learning, it is not a resolution approach for an FLP.

3.6.1.3 Reinforcement Learning

Our sample further yielded a small number of papers that could be attributed to Reinforcement Learning. We surprisingly found no publications that directly use Reinforcement Learning as a resolution technique. The remainder of this sample is discussed below.

Tarkesh et al. (2009) present an approach to FLP based on a multi-agent system in which the plant layout is generated by the interactions of the agents. Each agent corresponds to a production unit with inherent characteristics, emotions, and a certain amount of money, which together describe its utility function. The available credit of an agent is adjusted during the training phase, while each agent tries to adjust its utility function to minimise its total layout costs in competition with others. Since each agent's pricing decision has a strong influence on the pricing proposals of other agents in later decision iterations, and since each agent's pricing proposal is directly related to the last system state, this procedure can be considered a Markov chain and thus modelled as a Markov Decision Process (MDP). This modelling is not discussed in their paper. However, the iterative adjustments of the amount of money available to the agent are represented by a modified Q-Learning notation. Nonetheless, this paper does not satisfy our IC2, as it uses a multi-agent system as a resolution approach and is thus excluded from our Prio 1 pool.

In Kondoh et al. (2000), a self-organisation approach for cellular manufacturing systems based on Reinforcement Learning is proposed. The FLP is, in essence, represented as a quadratic assignment problem in the sense that cells are available for placing equipment on a grid. The reward is computed as travel time between cells and operation times, based on a certain operation sequence. An underlying MDP formulation cannot be distinguished as the notation differs from what is currently predominant in RL: in this paper, the action space is referred to as "tactics". A definition for the state space is not given. The authors further do not specify whether they employ any well-known RL paradigm, such as Q-Learning as above. However, the tactic selection probability strongly resembles typical epsilon-greediness from RL. Furthermore, it becomes clear that this FLP is designed as an episodic task, as the pallet unit moves back to the entrance after obtaining a reward. While this is an interesting early approach to RL in FLP, the goal in this paper is to choose one out of three different configurations (line, flexible, or island) for the manufacturing system. Layout planning, on the other hand, is explicitly mentioned as a future research topic, effectively putting this paper beyond our research scope (i.e. Prio 2).

The paper presented by Ono et al. (2001) uses tree search and a backtracking algorithm with time-reducing heuristics to make up a layout and schedule synchronously. Some characteristics closely resemble standard RL concepts: for rearrangement of the floor plan, the equipment can move along a grid (i.e. the action space is top, down, left, right, rotations, and idle). The reward, or penalty, is computed according to the number of workers needed to fulfil a given action. The number of turns taken in a simulation corresponds to the training epochs in RL. Plus, the mechanism to award bonus points for favourable layouts resembles rewards structures of MDPs, e.g. "if an action reduces the distance (...) add points to its weight". While time constraints point to an episodic design, spatial constraints, such as the freeness of collision, provide a hint to a finite state space. However, the construction of the tree to be searched is not specified, such that no concluding statements about the state space can be deduced. The search can occur in two modes: search alone or in cooperation with a user. In the latter mode, an interface enables users to give instructions to the system to improve the quality of the solution or search speed. Whereas the paper's authors note that in their system, "actions are decided one by one," which may satisfy the Markov property, their analysis shows that the system needs user help in problems larger than 20 objects, which corresponds approximately to the performance decay reported for exact approaches. As points like this lack the learning component critical to RL, we consider this paper a dynamic programming approach rather than an RL one and thus code it as such.

3.6.2 Discussion of Findings

Figure 3.10 below summarises the results of our analysis in absolute numbers per learning paradigm[1]. We found a total of nine publications belonging to Supervised Learning, none of which were used directly as the FLP resolution method (IC1). One explanation that stands to reason is that the lack of labelled data shift strongly drives their use as auxiliary functions or preparatory planning tasks. As Chung (1999a) argues, Facility Layout Problems are always ill-structured and their information is noisy, uncertain, or incomplete, making it hard to obtain a proper supervisor. The well-known drawback of FLPs is that their NP-hard nature prohibits optimal solutions at problem sizes larger than around 15. Layouts labelled through expert judgment are of a rather subjective quality and disqualify as suitable training data with respect to layout *optimisation*. The quality of approximating approaches is, per their definition, sub-optimal, albeit workable. Using layouts generated by GA, ACO, construction algorithms, or the like instead of training data might allow a Neural Network to generalise over a variety of feasible solutions. With Machine Learning methods being statistical tools (Russell and Norvig, 2016), it is debatable whether these abstractions could produce layouts with higher quality on their own rather than replicating approximated solutions. It remains to be seen whether SL-based resolution techniques will find their way into FLP research.

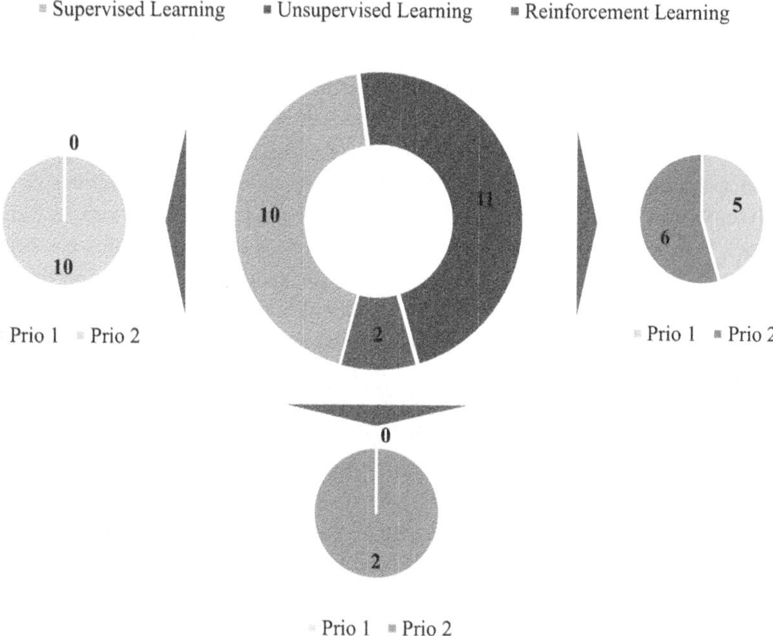

Fig. 3.10 Distribution of occurrence of learning paradigms in final dataset

We further identified 11 contributions from Unsupervised Learning, roughly half of which were used directly to solve FLPs. All but one relevant publications employ Kohonen or Hopfield Neural Networks. The underlying motivation for this appears to be a low computational load compared to meta-heuristics, especially for larger-sized problems, as well as faster and less labour-intensive implementation than the traditional (i.e. manual) Systematic Layout Planning method. The remaining papers deal with a major research stream named group technology, which can be considered an upstream planning process in production planning, which is not *primarily* concerned with layout design. Nonetheless, approaches using generative adversarial networks (GAN) in architectural research (Huang and Zheng, 2018) or variational autoencoders in (surface) layout designs in material science (Zhang and Ye, 2019) demonstrate ongoing research potential and might deserve further attention in FLP research.

Lastly, we found a low number of two papers using Reinforcement Learning approaches, yet only one pertinent to IC1. While we could identify evidence of reward or penalty formulations and actions space definitions, we did not find any cue for the use of Markov Decision Processes or the discussion of problem formulations satisfying the Markov property. Instead, problems were solved via exact approaches or multi-agent systems where the reward earning mechanism in Q-Learning merely played a minor role and was not further discussed. The one publication that most resembled an RL application did not have layout design as its focus (thus violating IC1), yet mentioned this application as future research scope. A forward search starting from this paper does indicate that this has been fulfilled to date. The pervasion of RL techniques, such as Q-Learning, in operational research is not overtly widespread, as another recent review investigated in which these techniques account for 1.5% of the papers (Weitzel and Glock, 2018). Apart from this, we observed that some of the methods proposed for FLPs originate from other practical problems. Yeh (2006) notes that SA has been effective in solving the travelling salesman problem (TSP), with respect to solution quality rather than to computing times. Similarly, Ateme-Nguema and Dao (2009) state that Hopfield networks had previously been used as TSP solutions. Given that Reinforcement Learning applications can be found in job-shop scheduling problem literature (Burggräf et al., 2020; Gabel and Riedmiller, 2012; Gambardella and Dorigo, 1995; Stricker et al., 2018), AGV routing (Malus et al., 2020) and have further been proposed as a future research topic on material handling system location planning (Lee et al., 2019), adopting RL as resolution technique for FLP, based on re-formulating classic layout formulations as Markov Decision Process, certainly stands to reason.

Figure 3.11 shows the temporal distribution of the papers mentioned above. One can observe that there was no research activity on the topic of ML in FLP before 1995, followed by a surge of activity until the year 2004. Contributions after this point in time are marginal. We attribute this to the end of the AI winter in 1997 (Russell and Norvig, 2016), which may have led to fresh funding for ML research for different practical domains. Pursuing this theory by analysing whether this pattern is mimicked by other disciplines goes beyond the scope of our review. Hence, we leave this observation to further speculation.

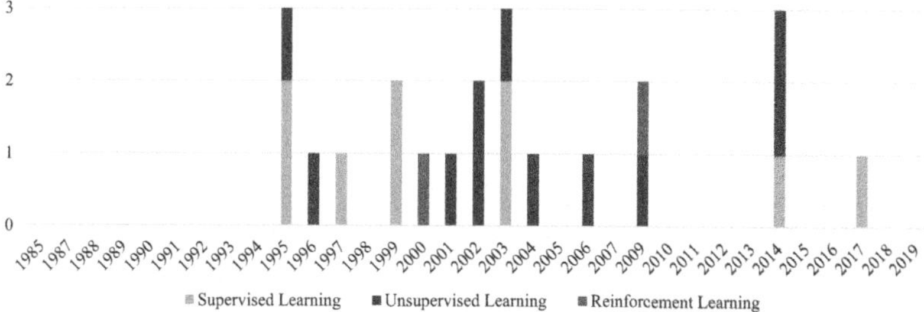

Fig. 3.11 Time series analysis of ML-related publications in FLP between 1985 and 2020

Interestingly, 86% (19 out of 22) of the papers examined make use of ANN for both supervised and unsupervised approaches. As these papers do not discuss their decision regarding other algorithms from Supervised Learning, as shown in Fig. 3.4, we cannot credibly conclude on the reasons. However, taking into account the research interest in ANN, exemplarily displayed for the database Sciencedirect in Fig. 3.12, we can observe that the peaks we see for the years 1995 and 2014 in Fig. 3.11 correlate with high relative changes in general research interest. Figure 3.12, on the other side, does not explain the peak in 1999 and merely provides a slight hint for the one in 2014. Yet, this leads us to the assumption that model selection was primarily influenced by zeitgeist rather than consideration of concurrent algorithmic alternatives.

In summary, we find that there is a distinct number of publications reporting on using ML techniques, mainly from the field of Unsupervised Learning, where said techniques are not directly employed to generate layouts, i.e. as resolution methods. These publications from the field of "group technology" aim at assigning machines to a certain number of "cells" with similar machines to minimise material flow crossings between cells. As such, these publications/techniques can be said to generate the internal contents of blocks in block layouts, but they do

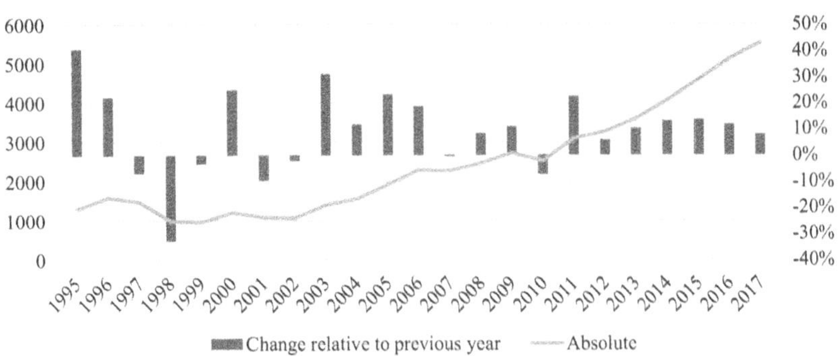

Fig. 3.12 Time series analysis of ANN-related papers on Sciencedirect between 1995 and 2017

not address the FLP in a way that determines the position of cells in a plant. This finding is in line with the analysis performed by Renzi et al. (2014) who indicated that AI techniques are primarily harnessed for grouping/cell formation and scheduling issues and not for layout design.

The results presented in the previous section have highlighted that intelligent approaches using any of the three Machine Learning paradigms Supervised, Unsupervised, and Reinforcement Learning plus the cross-sectional paradigm Deep Learning are under-represented in FLP literature relative to other approaches. Within our final sample used for our synthesis, we found a relatively even spread between Supervised and Unsupervised Learning techniques. However, their importance for FLP problems varies as we identified no paper where Supervised Learning is used for FLPs while we found five for Unsupervised Learning. Reinforcement Learning is even less represented, with a total of two papers, none of which satisfy both of our inclusion criteria.

3.6.3 Methodology

3.6.3.1 Systematic Review

We have conducted a review with a *sensitive* search strategy meaning that the goal is to identify as much relevant material as possible (Harden et al., 1999). We deliberately chose not to limit the search strategy to avoid the risk of missing relevant studies. Instead, we employed a clustering algorithm to aid us in coding papers. The results and implications of this are discussed hereinafter.

Our search yielded a raw dataset containing a total of 11,851 papers. Considering that we had 1,290 potentially relevant papers after all pre-processing and clean-up steps, we can ascertain our *precision* to be 0,11. In other words, before even entering the screening stages, we have a dropout rate of 89%, resulting in a significant upfront effort and thus a costly analysis. Concerning the selection of papers for the narrative synthesis, our review has a disproportionally low precision of 0.002 (24 in 11,851 papers). Our analysis regarding papers of which databases were still contained in the final study sample (see Fig. 3.13—OpenReview is excluded as it provided 0 results) shows that some of the databases we used suffer from a dropout rate next to 100%. This finding indicates that reducing the number of databases used is not necessarily with detriment to study quality but might increase precision and decrease review cost. As this result is always specific for the search and review study in question, our analysis does not warrant a generalisation of good practice. We nonetheless propose to invest a substantial amount of effort into the scoping search (Step 1 in the review framework used herein) to fine-tune the selection of databases.

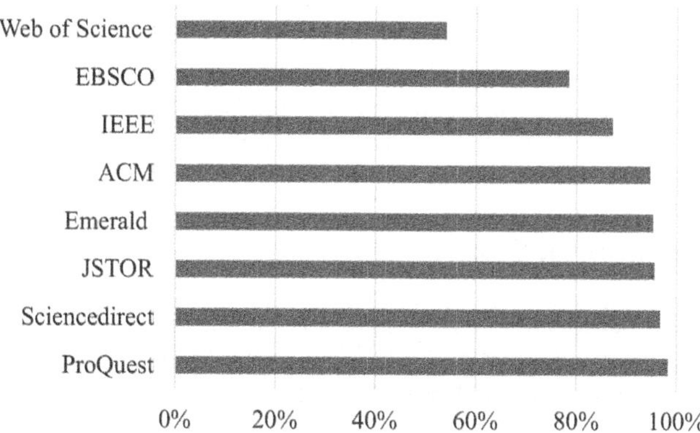

Fig. 3.13 Percentages of publications dropped per database along the review process before reaching screening stages

Our second proposition focuses on reducing losses in screening processes. We employed two distinct methods to support the analysis process: 1) a search algorithm to re-create title-abstract-keyword searches to filter out irrelevant contributions (these with facility layout problems merely being a side-notion) and b) a review team to increase inter-rater reliability. While any search automation is indispensable to sift through tens of thousands of references, special care must be exercised as to the results. We double-checked all exclusions suggested by the search algorithm and occasionally found contributions we considered relevant and thus re-included them manually. As a result of this multi-level filtering, there was still a body of 1,290 papers left to manually screen. We thus further recommend following the good practice as given by e.g. Gough et al. (2012) and to assemble a review team to reduce screening time and avoid bias in the analysis.

3.6.3.2 Clustering Algorithm

At the level of Stage 1 Screening, the clustering algorithm identified eight papers relevant to the research question while the manual screening found 49, yielding an accuracy of 14%. This number suffers from the fact that some papers used a different AI framework which considers meta-heuristics as an AI technique. Including these false positives increases the accuracy to 16%. Regarding the complete dataset, the algorithm correctly classified 185 out of 1,290 papers, likewise yielding an accuracy of 14%. Considering only the subset of papers relevant to the clusters, i.e. all papers manually assigned to GA, ACO, PSO, TS, SA, or AI, the classification rate increases to 37%. Based on these results, we conclude that using the clustering algorithm as a stand-alone technique is insufficient with respect to accuracy. However, this review can serve as an indicator for potential time-savings in lengthy review processes of 10–20%, depending on the level of trust put into the algorithm and thus the amount of rework required.

3.7 Section VI: Conclusion

Our review set out to analyse the extent to which Machine Learning techniques, herein defined as a subset of Artificial Intelligence, have been used in scientific literature as resolution approach for Facility Layout Problems.

Through a systematic literature review, we collected a set of 1,290 contributions in a sample of 11,851 papers retrieved from nine different databases plus another 134 contributions from snowball sampling. Analysing these contributions with respect to two inclusion criteria, i.e. the relevance to the field of FLP (IC1) and reports of using an ML technique as a resolution approach to solving FLP (IC2), yielded only 23 contributions that had a distinct AI focus. Two more papers were included as they showed traits pertinent to ML techniques.

As the essence of our in-depth full-text analysis of 25 papers, we can answer our research question, *How have different Machine Learning algorithms been used as resolution techniques for Facility Layout Problems?*, as follows: while there is some evidence for all three ML learning paradigms (Supervised, Unsupervised and Reinforcement Learning), their focus on FLP resolution approaches is scarce.

None of the papers found using SL were relevant to FLP resolution approaches. Instead, SL was primarily used as an adjunct optimiser for other resolution approaches or employed in planning tasks preceding that of layout design. We infer that the under-representation of SL could be driven by a general lack of structured, labelled data.

For UL, we identified the most prominent algorithms to be SOM and Hopfield Networks, respectively. Their use is motivated by a lower computational load for large problems than is required for similar meta-heuristics. However, we also detect a downswing in research interest in UL since a peak in the early 2000 s.

Additionally, we analysed that despite a wide variety of available ML algorithms, ANN were strikingly prominent. As a consequence, following SOM, backpropagation networks rank second in terms of occurrence. As the papers visited do not provide unambiguous evidence for the selection of ANN over other alternatives, using a time series analysis of research interest in ANN, we infer that the choice of ANN was likely to be a result of the spirit of the time rather than the ineptness of other algorithms.

Lastly, we found only two contributions in RL yet none of these satisfied IC1 and IC2. This finding indicates that RL has not yet found its way into FLP as a resolution approach. This is remarkable, as a distinct number of studies make use of simulations, requiring defined environments which could also be used to model MDPs.

In summary, this analysis contributes to the body of knowledge on the topic of facility layout problems by refining previous taxonomies. Specifically, to date, there existed no classification that decomposed intelligent approaches into the branches symbolic and sub-symbolic as well as the Machine Learning paradigms and associated algorithms. These updates allow future researchers to classify their work in more detail and encourage comparative studies on the effectiveness and efficiency of alternative algorithms on certain problem sets. We further contribute by highlighting white spaces, such as the non-existence of several algorithms in our classification in FLP to date. Thus, as future research work, we encourage the investigation of both SL and RL approaches. For SL, deep learning-based solutions may prove to be valid topics. Deep Learning models trained on known layout alternatives may be able to predict or rate new layouts simply by visual input, such as raw images, by being able to identify certain objects in the layout. By leveraging computer vision techniques, this could become an opportunity for on-the-fly layout evaluation in group decision-making settings. The idea to use RL as a resolution approach in future FLP research draws from insights that are being generated in other complex problem domains in production research. Research work in the area could include formal MDP notations as a foundation for using RL, insights on the modelling prerequisites to make RL algorithms trainable, or studies on the usability of various available algorithms (e.g. when will tabular Q-Learning approaches reach their limits? Are there differences in convergence or training speeds for different algorithms, for instance, those using discrete action spaces compared to those with continuous actions spaces).

3.8 Declaration of Own Contribution

All the work presented for the systematic research was carried out by the doctoral candidate. This includes the definition of the review scope, the definition of the search string and the databases, the definition of quality criteria and research questions, the search of the literature corresponding to the search string, the evaluation of the literature according to the quality criteria, the development of the framework and the classification of the literature into this framework, the analysis of the literature by means of descriptive statistics as well as the subsequent synthesis and critical reflection of the literature. In addition, manuscript drafting was performed by the doctoral candidate. Both co-authors, Prof. Dr. Burggräf and Dr. Wagner, contributed ideas to the research concept and the internal review process.

Publication II: Gym-Flp: A Python Package for Training Reinforcement Learning Algorithms on Facility Layout Problems 4

4.1 Abstract

Reinforcement learning (RL) algorithms have proven to be useful tools for combinatorial optimisation. However, they are still underutilised in facility layout problems (FLPs). At the same time, RL research relies on standardised benchmarks such as the Arcade Learning Environment. To address these issues, we present an open-source Python package (gym-flp) that utilises the OpenAI Gym toolkit, specifically designed for developing and comparing RL algorithms. The package offers one discrete and three continuous problem representation environments with customisable state and action spaces. In addition, the package provides 138 discrete and 61 continuous problems commonly used in FLP literature and supports submitting custom problem sets. The user can choose between numerical and visual output of observations, depending on the RL approach being used. The package aims to facilitate experimentation with different algorithms in a reproducible manner and advance RL use in factory planning.

Heinbach, B.; Burggräf, P.; and Wagner, J. (2024) "gym-flp: A Python Package for Training Reinforcement Learning Algorithms on Facility Layout Problems". In: *Operations Research Forum* 5, 20. https://doi.org/10.1007/s43069-024-00301-3.

B. Heinbach, *Reinforcement Learning-Based Planning of Factory Layouts*, Findings from Production Management Research ,
https://doi.org/10.1007/978-3-658-51554-6_4

4.2 Section I: Introduction

Facility layout problems (FLP) are an important class of optimisation problems in operations research (OR). The FLP constitutes a fundamental aspect of operations research and industrial engineering, addressing the strategic arrangement of components within a facility to optimise efficiency, reduce costs, and amplify productivity. The optimal layout design significantly impacts material flow, transportation costs, and worker performance (Drira et al., 2007).

The FLP is NP-hard and has been tackled using an extensive array of modelling and resolution techniques (Burggräf, Wagner, Heinbach, 2021; Drira et al., 2007; Hosseini-Nasab et al., 2018; Pérez-Gosende et al., 2021). In recent years, the use of Machine Learning (ML) techniques has experienced a surge for other important tasks in manufacturing, e.g. (Burggräf et al., 2018; Burggräf et al., 2019; Burggräf, Wagner, Koke, Bamberg, 2020; Burggräf, Wagner, Koke, Steinberg, 2020; Dogan and Birant, 2021; Wuest et al., 2016). Special consideration has been conveyed to Reinforcement Learning (RL), a specific sub-class of ML, in manufacturing and supply chain research (Bahrpeyma and Reichelt, 2022; del Real Torres et al., 2022; Panzer and Bender, 2022; Rolf et al., 2023). For combinatorial optimisation problems, provided sufficient training resources and a supervised initialisation, RL approaches bear the potential to generalise on new instances if the training was performed on a small distribution of the problem to exploit its structure (Bengio et al., 2021).

Building upon these insights, we infer that facility layout planning systems based on RL will achieve a level of proficiency comparable to that of human layout planners. A fundamental premise to consider is that the distinctive attributes of DRL, such as the potent feature extraction and function approximation capabilities offered by Convolutional Neural Networks (CNN), can potentially be harnessed and extended to the realm of facility layout planning. Thus, when being trained on a sufficiently large distribution of FLPs, the definition of which is yet to be made, trained DRL agents might be able to make proposals for factory layouts while eliminating the need for training new RL agents and modelling problems, thus facilitating and speeding up facility layout planning.

To address this, we introduce an open-source Python package providing a standardised interface that allows researchers to leverage commonly used FLP benchmark problems for training and testing RL algorithms in a comparable and reproducible manner. Its purpose is not to compete with existing heuristics in FLP research but to provide a curated set of both discrete and continuous well-studied FLP problems known in literature to decrease the up-front modelling effort and to make it accessible to an FLP research community interested in investigating the practical utility of RL approaches in facility layout design.

The contribution of this paper is threefold: we service the FLP community with a toolbox that (1) provides open-sourced environments for well-known benchmark FLP problems and (2) presents an interface to these environments to reduce implementation efforts, so as to (3) enable comprehensive benchmarking of FLP implementations against readily available RL algorithms. Thus, we hope to assist in showing whether RL can prove to be useful for FLP research and practitioners.

4.3 Section II: Facility Layout Problems

The central objective of Facility Layout Problems involves determining the optimal spatial arrangement of workstations, departments, machines, or equipment within a facility. This arrangement profoundly influences factors such as material handling expenses, production time, worker movement, and overall facility utilisation. FLPs encompass identifying the most favourable positioning of these components while adhering to constraints such as space limitations, safety regulations, and workflow prerequisites (Tompkins et al., 2010; Zuniga et al., 2020). Typically, one central element of an FLP is to minimise the transport intensity between facilities, given their pairwise flow relationships and distances. This transport intensity is commonly referred to as Material Handling Cost (MHC). Thus, the general optimisation objective of an FLP regarding the MHC is given as follows:

$$minimize : MHC = \sum_{i}^{n} \sum_{\substack{j \\ i<j}}^{n} f_{ij} d_{ij} c_{ij} \tag{4.1}$$

where f_{ij} denotes the flow relationship between facilities i and j, d_{ij} is the distance between i and j, usually measured from their centre point, and c_{ij} is a scaling factor to account for different modes of transportation in between facilities.

A central aspect of solving Facility Layout Problems revolves around the mathematical formulation of the problem itself, as the choice of formulation governs the complexity of the solution. FLPs can be classified into discrete and continuous formulations based on the treatment of spatial arrangement variables. In *discrete* FLPs, the space is divided into a finite number of predefined locations, often termed grid points or nodes. Each component or workstation is assigned to one of these discrete locations. In *continuous* FLPs, the placement of components is treated as a continuous variable, allowing flexibility in layout design. Components can be placed at any point in the continuous space (Drira et al., 2007).

Discrete FLPs are often modelled as a Quadratic Assignment Problem (QAP). The QAP deals with assigning a set of facilities to a set of locations in a way that minimises the sum of weighted distances between facility pairs (Koopmans and Beckmann, 1957). This combinatorial optimisation problem finds applications in fields ranging from manufacturing to telecommunications.

Continuous FLPs, in turn, are modelled with various degrees of freedom: the Flexible Bay Structure (FBS) notation has been created by Tong (1991). It allows the departments to be located only in parallel bays with varying widths. Bays are bounded by straight aisles on both sides, and departments are not allowed to span over multiple bays (Konak et al., 2006). The width of the bays is determined by the cumulated area demand of the facilities assigned to each one using the equation below (Ulutas and Kulturel-Konak, 2012). The Slicing Tree Structure (STS) decomposes irregular shapes into simpler rectangles, easing the layout process. It involves iteratively partitioning a larger rectangular area into smaller rectangles through horizontal and vertical cuts. This method aims to find an arrangement of smaller rectangles that optimises the given layout objective (Tam, 1992). Lastly, the Open Field Layout Problem (OFP) involves designing the layout for facilities with vast open spaces, such as warehouses or large manufacturing areas. The objective is to optimise material movement and accessibility without the restrictions or constraints that would be imposed by such arrangements as a single row or loop layout. Instead, a key concern of the OFP is that facilities are to be arranged free of overlaps (Niroomand et al., 2015).

Historically, various strategies have been employed to tackle Facility Layout Problems, reflecting diverse problem formulations and solution methods (Hosseini-Nasab et al., 2018). Classical techniques, such as the systematic layout planning (SLP) approach, emphasise human factors, expertise, and experience to create efficient layouts. In the SLP approach, qualitative and quantitative considerations are combined to generate layout alternatives. However, such methods might fall short in complex environments with numerous variables and constraints (Yang et al., 2000).

Other prominent approaches are mathematical optimisation techniques like integer programming, genetic algorithms, simulated annealing, and particle swarm optimisation. These methods leverage computational power to systematically search for optimal or near-optimal solutions, taking into account various constraints and objectives simultaneously. These algorithms have proven effective in generating layouts that minimise transportation costs, minimise material handling time, and maximise facility throughput (Anjos and Vieira, 2021).

More recent approaches make use of ML tools. Despite some evidence of using Artificial Neural Networks in the facility planning approach, see for instance (García-Hernández et al., 2014; Tsuchiya et al., 1996; Ueda et al., 2002), a recent study has shown that ML methods, and RL specifically, have seen comparably little use in FLP research (Burggräf, Wagner, Heinbach, 2021). This is remarkable since RL has been employed to solve several real-life industry-related problems, such as electric energy storage systems (Weitzel and Glock, 2018), job-shop scheduling (Burggräf et al., 2018; Burggräf et al., 2020; Kuhnle et al., 2019; Shi et al., 2020), autonomous guided vehicle routing (Malus et al., 2020), or the travelling salesman problem (Khalil et al., 2017). Contrary to this lack of application evidence, recent work (Heinbach et al., 2023; Ikeda et al., 2022; Klar et al., 2021, 2023; Klar, Hussong et al., 2022; Unger et al., 2024; Unger and Börner, 2021) demonstrates the need for and interest in the application of RL to FLPs.

4.4 Section III: Reinforcement Learning

Reinforcement Learning has gained prominence due to significant research advancements in creating learning agents capable of excelling in virtual arcade game environments (Mnih et al., 2015). RL, most notably its deep learning variant for higher-order problems, Deep Reinforcement Learning (DRL), has been demonstrated to outperform expert human players in board games (Silver et al., 2016) and computer games by learning merely from visual input (Mnih et al., 2015).

In RL, an agent learns to maximise a reward signal by exploratively interacting with its environment. The goal of the agent is to learn a policy that maps states of the environment to actions in such a way as to maximise the expected discounted cumulative rewards over time:

$$E_{\pi}[R_t] = E_{\pi}\left[\sum_{i=1}^{\infty} \gamma^i r_{t+1}\right], \tag{4.2}$$

where $\gamma \in (0, 1]$ is a discount factor that governs whether rewards at later training progress are treated myopic or farsighted.

A reinforcement learning problem can be formally defined as a Markov decision process (MDP), which mathematically translates problems into a sequential decision-making process. An MDP consists of:

- A set of states, S, that represent the possible configurations of the environment.
- A set of actions, A, that the agent can take in each state.
- A reward function, $R : SxA \to R$, which assigns a reward to each state-action pair.
- A transition function, $T : SxA \to S$, which specifies the next state that results from taking a given action in a given state.

The agent begins in an initial state and at each time step t, selects an action a according to a policy π and transitions to a new state. The reward at each time step is determined by the state-action pair and the resulting state is determined by the transition function (Sutton and Barto, 2018). A representation of the MDP is shown in Fig. 4.1.

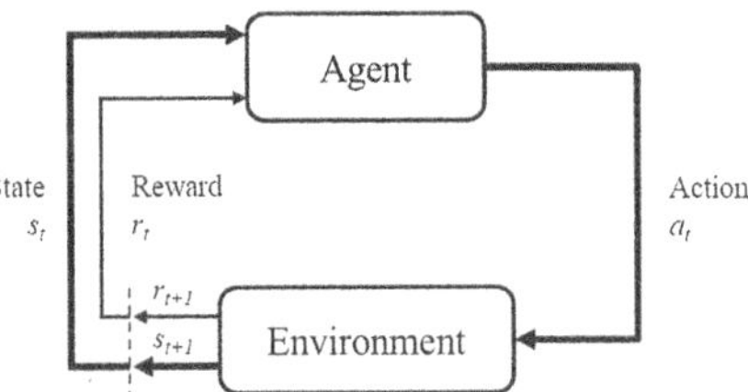

Fig. 4.1 Representation of the MDP

In reinforcement learning, there are two main approaches to learning a policy: model-based and model-free.

Model-based RL involves learning a model of the environment, which allows the agent to make predictions about the consequences of its actions. With this model, the agent can plan its actions using a search algorithm such as value iteration or policy iteration. The advantage of model-based RL is that it can learn an optimal policy more efficiently since it can use its model to predict the outcomes of actions and plan accordingly. However, it can be difficult to learn an accurate model of the environment, especially if the environment is complex or the state space is large. Model-free RL, on the other hand, does not involve learning a model of the environment. Instead, the agent directly learns a policy by interacting with the environment and receiving rewards. Model-free reinforcement learning is simpler to implement and can learn directly from raw sensory data, but it can take longer to learn an optimal policy compared to model-based methods (Sutton and Barto, 2018).

Two other fundamental approaches within RL are value-based and policy-based methods. Value-based methods focus on estimating the value of taking various actions in different states. They aim to learn a value function, such as the Q-function in Q-learning, which assigns a value to each state-action pair. Agents then make decisions by selecting actions with the highest estimated value. In contrast, policy-based methods aim to directly learn the optimal policy, which is a mapping from states to actions. Instead of estimating the value of actions, policy-based methods adjust the policy itself to maximise the expected cumulative reward (Sutton and Barto, 2018).

Furthermore, reinforcement learning algorithms can be categorised into on-policy and off-policy methods. On-policy algorithms learn and improve the policy they currently follow. This means that the data used for learning must be collected using the current policy, which can limit exploration. Off-policy algorithms, on the other hand, allow an agent to learn from data generated by a different policy, enabling more efficient exploration and potentially better sample efficiency. Popular examples of off-policy algorithms include Q-learning and off-policy actor-critic methods (Dong et al., 2020).

4.5 Section IV: Leveraging Reinforcement Learning for Facility Layout Problems

In the contemporary landscape of optimisation challenges, the paradigm of RL emerges as a compelling alternative to tackle the intricacies of FLPs. With an inherent capacity to learn, adapt, and optimise over time, RL brings to the fore a dynamic approach that resonates with the multi-faceted nature of FLPs.

A hallmark of RL is its intrinsic inclination toward exploration and exploitation. This trait holds profound relevance in the context of FLPs, where sub-optimal solutions might often be disguised as local optima. RL algorithms can navigate this challenge by leveraging exploration mechanisms, thereby unearthing novel layout configurations that can lead to substantial performance enhancements. Adaptive heuristics embedded within RL techniques continuously refine strategies, ensuring a balance between tried-and-tested methods and novel explorations.

Yet another distinctive characteristic of RL, more precisely DRL, is the utilisation of Convolutional Neural Networks (CNNs) as described by Mnih et al. (2013). CNNs excel at extracting spatial features from structured data. FLPs involve spatial configurations of various components, such as layout maps or images depicting the facility. By integrating CNNs into the Reinforcement Learning framework for FLPs, CNNs can learn to generalise spatial patterns from one facility layout to others, enhancing the transferability of learned policies. This can be particularly valuable in scenarios where similar layout structures occur across different facilities.

One key challenge to be addressed is to design the problem representations and underlying Markov Decision Processes (MDP) in a way that they can accommodate different sorts of problem types and become less sensitive to problem sizes. By submitting fixed-size input information to the RL agent, i.e. an image with the same amount of pixels in both directions regardless of the number of facilities, one is able to assume control over the state space dimensionality.

The underlying hypothesis driving the development of this software tool is that the synergy between spatial feature extraction and policy optimisation can lead to more accurate, adaptive, and efficient solutions. CNNs empower RL agents to navigate the complexities of spatial arrangement while accounting for visual cues and correlations inherent in facility layouts.

The Python package presented herein aims to achieve this synergy by providing a visual representation of the FLP to the RL agent. We do so by encoding the flow intensity matrix, both row and colour-wise, in the colour channels of an image. The flow intensity matrix contains information on both the flows and distances from Eq. (4.2), where rows can be interpreted as flow sources and columns as sinks, respectively. By encoding the flow intensity information, the RL algorithm should be able to abstract structural information of the problem. That is, the machine with the lowest transport flows both in- and out-bound will be shown as dim red (e.g. RGB = (40, 0, 0)) whereas machines with higher transport relationships will move closer to purple or even white ((40, 255, 255) or (255, 255, 255)). Figure 4.2 visualises this concept, the details of which are outlined in Sect. 4.6.3. This functionality, coupled with the respective reward assignment, should enable an RL agent to—at least—reproduce basic heuristics, such as 'place the machine with highest flows in the centre and arrange all others around it', similar to the triangular placement method of Schmigalla (1970).

As Li et al. (2023) have pointed out, a key objective for using DRL in manufacturing research and practice is to reduce manual involvement. With facility layout planning typically being a highly manual process, the exceptional representation learning capabilities of CNNs combined with the adaptive heuristics inherent to RL techniques can position them as a transformative force in contemporary FLP research.

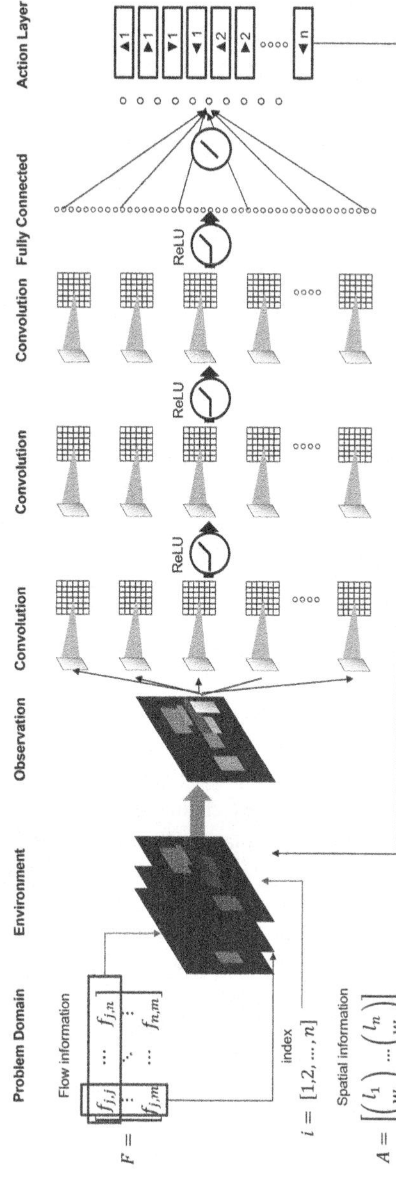

Fig. 4.2 Illustration of the concept of embedding FLP information in visual representations to make them digestible by Convolutional Neural Networks. Image adapted from Mnih et al. (2015) and Patel et al. (2019)

4.6 Section V: Tying it Together: The Gym-Flp Library for Using RL on FLPs

4.6.1 The Backbone: OpenAI Gym

In this section, we introduce the Python package gym-flp that can be used to train Reinforcement Learning agents on common or custom FLP instances. This software library is available at https://github.com/BTHeinbach/gym-flp.

The library we propose in this work is based on OpenAI Gym, a toolkit designed for developing and comparing RL algorithms. Gym provides researchers with access to benchmarks for training agents in the form of *environments* (Brockman et al., 2016). These environments include control problems, arcade games, and robotics. The availability of such benchmarks has played a significant role in advancing the field of RL (Hein et al., 2017). In fact, open-source libraries in general play a vital role in evaluating innovations in the field of algorithms using benchmarks on a variety of problems (Serra and O'Neil, 2020).

While originally developed for RL research, there are several examples of how OpenAI Gym has been used to address real-world research questions, such as controlling power systems on grid (Li and Du, 2018; Vázquez-Canteli et al., 2019) or on building level (Spangher et al., 2020), job-shop scheduling problems (Waschneck et al., 2018), simulated robotic motion (e.g. pick-and-place operations) (Zamora et al., 2016), industrial process control (e.g. valve settings for gas turbines or pitch angles and rotor speeds for wind turbines) (Hein et al., 2017), or communication technology optimisation (Gawłowicz and Zubow, 2018).

The potential for OpenAI Gym in OR has also been recognised: Hubbs et al. (2020) published OR-Gym, a Gym-based library that features the common OR problem types knapsack, bin packing, supply chain, vehicle routing, news vendor, portfolio optimisation, and travelling salesman problems. Similar to the contribution presented herein, they aim to encourage further development and integration of RL into optimisation and the OR community while also opening the RL community to many of the problems and challenges that the OR community has been wrestling with for decades. Nonetheless, despite being a special problem in OR research, their library does not support FLPs.

The examples given above provide substantial evidence that OpenAI Gym is a prudent choice for creating frameworks for RL usage in FLP research. For our intended purposes, we reversed the Gym logic by putting the development focus on the environments rather than the algorithms by encouraging the use of collections of common RL algorithms, such as Stable Baselines (Raffin et al., 2019) or Ray (Moritz et al., 2018). Effectively, we assert that this can speed up

RL research on FLPs since, as a first step, researchers can resort to selecting a resolution technique from a well-established suite of algorithms to use on their problems before having to develop new or amend existing algorithms.

4.6.2 Package Structure

The implemented environments in this work are Python classes that inherit from the OpenAI Gym base class gymEnv and follow the same conventions regarding required class methods. This is because certain RL algorithm frameworks include a method for checking that the inputs and outputs of the environment are compatible with the agents. Figure 4.3 shows the outline of the package topography, the explanations of which follow thereafter.

4.6.3 Implementation Details

For an FLP to be solved using RL, its components need to be translated into the key concepts of an MDP as introduced in Sect. 4.4.

4.6.3.1 Observation Spaces

There are two modes for designing the observation space: RGB_ARRAY and HUMAN. The mode can be specified when initialising the environment.

In 'human' mode, the observation for a discrete Facility Location Problem (FLP) is a permutation vector of size n, sampled randomly without replacement, where n is the problem size of the instance. The permutation vector represents the available locations (numbered consecutively) and the assigned machines. The assumption is that every facility can be assigned to any location. Further restrictions can be implemented by those interested in this functionality.

For continuous FLPs, the observation includes centre point coordinates (x, y), , length (l), and width (w) information for each facility in $i = 1 \ldots n$, resulting in an observation space of $4 * n$. The range of possible values for an observation is determined using the Box space and is based on problem instance information in the *__init__* function.

One of the main motivations of the authors is to teach RL algorithms to rearrange factory plants using visual input alone, similar to the early advances in RL research on arcade games (Mnih et al., 2013). To facilitate this, we provide the rgb_array mode, in which observations are treated as images with dimensions corresponding to the sizes of the plants. The allowed values for the third dimension

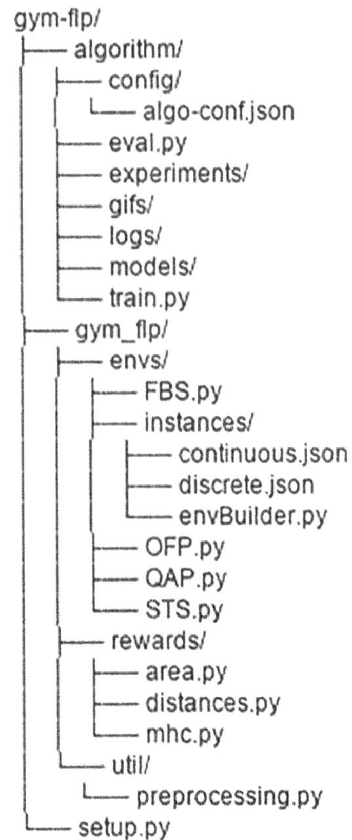

Fig. 4.3 gym-flp package structure. Module init files were omitted

are limited to the range [0, 255], and the observation is normalised (divided by 255 to have values in [0, 1]) when using Convolutional Neural Network (CNN) policies. The baseline algorithms implemented in Stable-Baselines3 and RLlib use predefined filter sizes, so the plant sizes may need to be adjusted accordingly on occasion. For example, the *CnnPolicy* in Stable-Baselines3 requires observed images to be at least 36 by 36 pixels in size.

To facilitate feature extraction in CNN-based policies and make all facilities on the plane identifiable, we encode certain information in the RGB colour space of the observation images. The index i of a facility is encoded on the red channel, while the row sums of the flow matrix F (representing how much a facility is a flow source) are embedded in the green channel and the column sums of F (i.e.

the sinks) are written into the blue channel. The underlying idea is that facilities with higher green and blue values have stronger flow relationships and may be located at the centre of all facilities, potentially allowing an RL algorithm to learn this rule through intuition or heuristics.

Figure 4.4 shows a visual representation of the rgb_array observations for each environment.

Note that the observation mode has an impact on the policy used by the algorithm. For instance, the use of human mode forbids using CnnPolicy as convolutional filters do not work on one-dimensional arrays.

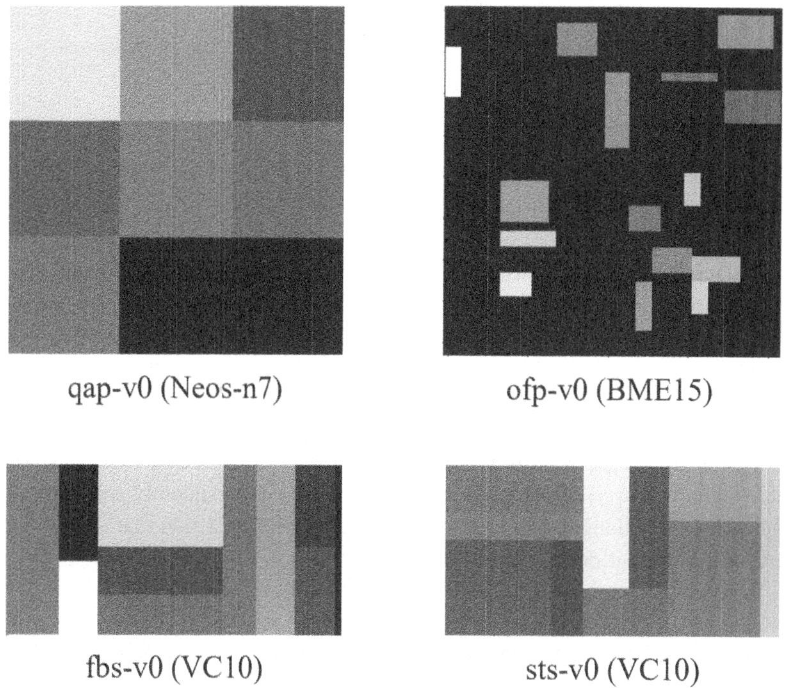

Fig. 4.4 Examples for environment renderings. The text in brackets describes the instance used

4.6.3.2 Action Spaces

Defining proper action spaces is likely to be as crucial in RL as reward engineering. This section explains the predefined implementations, yet we encourage all interested researchers to amend the actions for their respective purposes.

Aside from Box used for the observation spaces, gym.Spaces provides several other types of spaces, e.g., Discrete, MultiDiscrete, Dict, Tuple, Graph, MultiBinary, Sequence, and Text. To define the possible range of actions, gym-flp currently uses Box, Discrete, and MultiDiscrete.

Table 4.1 which spaces are supported in the respective environments.

Table 4.1 Summary of currently supported action spaces per environment

Environment	Discrete	Multi-Discrete	Box	Box (simultaneous)
qap-v0	✓	–	–	–
fbs-v0	✓	–	–	–
ofp-v0	✓	✓	✓	✓
sts-v0	✓	–	–	–

In qap-v0, the action space represents a pairwise exchange mechanism, where two facilities i and j will swap their currently assigned locations. We also supply a swap (i, j) to provide an action to remain in the current state. Since the swap (i, j) is identical to (j, i), the size of the action space is given as $\left(n^2 - n\right) * 0.5 + 1$.

fbs-v0 incorporates five discrete actions: *Permute* swaps two random positions in permutation vector, *Bit Swap* flips the value (0/1) of one random element in the bay break vector. With *Bay Exchange*, two randomly selected bays exchange their facilities. *Inverse* inverts the order of facilities in a randomly selected bay, and *Idle* does nothing, again to provide the agent with an action that allows it to remain in a state it deems optimal.

sts-v0 uses five discrete actions as well. In addition to *Permute* and *Idle*, the action space uses *Slice Swap* (swap two random positions in slicing order), *Bit Swap* (change *slicing* orientation at random position) and *Shuffle* (Create new random slicing order).

To date, the action spaces for ofp-v0 are the most comprehensive. The Discrete variant is a 1-D array of size $4 * n + 1$, where each facility can take any step in the north, east, south, or west direction at a step size that defaults to 1 but can be passed differently upon initialisation. The single additional action is again used as a way to retain the current state. The MultiDiscrete action space folds the four actions into a 2-D array. Here, all facilities are moved simultaneously in any of the four directions. The idle action is appended as the fifth option to every dimension. Lastly, there are two options using a Box space. Both of them use Cartesian coordinates on the factory plant ranging from point $(0, 0)$ to (Y, X), with Y and X corresponding to the plant width and length, respectively. The Box spaces differ in that facilities can be moved sequentially or simultaneously. This decision is passed upon initialisation using the boolean argument multi.

4.6.3.3 Reward Computation

The reward engine comprises up to three different components, depending on the environment and action space used. Since QAP, FBS, and STS are collision-free and within defined plant bounds, these environments only handle material handling cost (MHC).

The flows between facilities are stored in the package. The distance metric for continuous problems can be set upon initialisation with the options rectilinear, Euclidean and squared-Euclidean, where rectilinear is the default value. MHC computation is invoked as an instance of the sub-module *rewards.mhc*.

An intricacy to consider is that common RL approaches attempt to maximise cumulative reward, whereas the goal in FLP is to minimise transport cost. At the same time, the reward signal should enable generalisation and avoid reward gaming (i.e. the exploitation of an unintended loop-hole) as described in Unger and Börner (2021). To address these points, we have set up the reward component for MHC as a moving target. In every step, we compare the MHC at time step t $MHC_{s(t+1)}$ with the best-known MHC MHC_{best} of the current episode. A reward of 1 is assigned if the new MHC is lower, or 0 otherwise, see Eq. (4.3). Then, MHC_{best} is overwritten accordingly.

$$r_{MHC} = \begin{cases} 1, if MHC_{s(t+1)} < MHC_{best} \\ 0, otherwise \end{cases} \tag{4.3}$$

For the OFP environment, layout feasibility needs to be taken into account. In contrast to the environments where locations are determined, OFP needs to incorporate non-overlapping constraints. We, therefore, define the penalty term $p_{collision}$ (see Eq. (4.4)) that collects a penalty of 1 if one facility intersects the union of the remaining ones. We choose a value of 2 to prevent a sparse reward signal for actions that improved MHC (+1) but resulted in a collision.

$$p_{collision} = \begin{cases} 2, if F_i \cap \left(\bigcup_{j=1}^{i-1} F_j \cup \bigcup_{j=i+1}^{n} F_j \right) \neq 0 \\ 0, otherwise \end{cases} \tag{4.4}$$

Finally, the variants that rearrange facilities with a movement with a given step size (Discrete and MultiDiscrete) are at risk of moving facilities beyond the plant boundaries resulting in entering a state beyond the state space. Such actions are penalised in two ways: the current episode is terminated (similar to a game-over situation in arcade games) and a penalty $p_{off-grid}$ of 10 is assigned so as to overrule possible positive rewards from MHC and to be free of collisions, see Eq. (4.5).

$$p_{off-grid} = \begin{cases} 10, if s_{t+1} \notin S \\ 0, otherwise \end{cases} \tag{4.5}$$

In consequence, the total reward per step (in OFP) is given as:

$$r = r_{MHC} - p_{collision} - p_{off-grid} \tag{4.6}$$

Reward engineering is said to be one of the most important tasks in RL research. Therefore, all researchers using gym-flp are encouraged to redefine the reward computation according to their respective needs and suitable for their problems in question. For instance, additional reward or penalty terms can be defined inside gym-flp.

4.6.3.4 Agent

This package uses Stable Baselines 3 (Raffin et al., 2019) as the framework for RL and the algorithms it currently supports (as of version 1.8.0). Thus, the algorithms available off-the-shelf in gym-flp are

- Deep Q-Networks (DQN) (Mnih et al., 2013)
- Proximal Policy Optimisation (PPO) (Schulman et al., 2017)
- Advantage Actor Critic (A2C) (Mnih et al., 2016)
- Deep Deterministic Policy Gradients (DDPG) (Lillicrap et al., 2015)
- Twin-Delayed DDPG (TD3) (Fujimoto et al., 2018), and
- Soft Actor-Critic (SAC) (Haarnoja et al., 2018)

Prior to training, hyper-parameters for the algorithm are loaded from the algo-conf.json file located inside./algorithm/config/. The hyper-parameters are the baseline values as stored in the Stable Baselines 3 classes. These can be altered by

a) Manually changing the parameters in the configuration file by navigating to the source folder
b) Passing the parameter values along with train.py, see Sect. 4.7.2

When choosing option b), parameters can be passed as a single value for one-off training (see Sect. 4.7.2.4) or as two values describing the lower and upper bound for an optimisation run (see Sect. 4.7.2.5).

4.6.3.5 Environment

The environment is the heart of the package and contains the state transition dynamics for the observation and action spaces defined above. The environment follows the Gym convention and thus contains the functions __init__(), reset(), step(), and render().

Environments in Gym are typically *episodic*, i.e. they usually possess a terminal state that aborts an episode and resets the environment. One such terminal state in the context of FLPs could be the optimal state of the layout given the objective function. However, the implementation assumes that no optimum is known a priori. Therefore, the gym-flp environments are designed as *continuous* tasks. To nonetheless enable the use of episode-based callbacks and evaluations, episodic behaviour needs to be mimicked. We have therefore included the following termination criteria:

- An action leads to a state outside of the state space (applies mostly to OFP)
- No improvement of MHC has been made in a number of consecutive steps

The first criterion prevents the agent from learning actions that produce infeasible layouts and are penalised accordingly, see Sect. 4.6.3.3. The second criterion assumes that the agent has reached a region close to a (local) optimum.

4.7 Section VI: Usage

This section provides basic examples of how to use the described package with Reinforcement Learning approaches. Following some installation hints, we provide execution commands for training and tuning scripts provided with the package.

4.7.1 Installation

The procedures below were tested on a Windows 10 OS and assume the usage of an Anaconda, Pycharm or Visual Studio Code distribution. gym-flp requires Python with a version starting from 3.7. The following ways exist to install the package.

4.7.1.1 PyPI Installation

Gym-flp supports PyPI installation which is the most straightforward means of installation. The package can be conveniently installed by opening a terminal window in the Python home directory and running the command:

pip install gym-flp

The use of virtual environments, such as virtualenv, conda or pipenv is highly encouraged.

4.7.1.2 Installation from GitHub

The next option is to directly install the package from the GitHub source. This can be achieved by opening a command terminal and running the command:

pip install git + git://github.com/BTHeinbach/gym-flp.git

4.7.1.3 Cloning GitHub Repository

If all else fails, navigate to the GitHub repository under https://github.com/BTHeinbach/gym-flp and download the.zip package to a destination folder of your choosing. Next, open up a terminal, navigate to the destination folder using the command cd and type the command below (Note the full-stop after the space). This will install the package in editable development mode, allowing easier changes to it.

pip install -e.

The examples below demonstrate the usage and results.

4.7.2 Basic Experiment Workflow

According to the experiment workflow presented in Fig. 4.5, in the current version of gym-flp, researchers have the following experimental design options:

- Common problem instance or custom problem
- One-off or optimisation
- Meta-Study or algorithm tuning

The core of the package are the 'train' and 'evaluate' blocks. Train will call evaluate from within to test the agent on the new instance of the environment. But both can be called individually from a command line interface. Calling 'evaluate' independently can be useful to test an agent with different random seeds.

The training scripts will create a model according to the script name and the arguments passed. Finally, when training is completed, the program will run one episode in the environment until termination and log reward and MHC per step for evaluation. The Tensorboard logs, final and (if applicable) intermediate models, experiment JSON files, and GIFs from the evaluation run are stored in respective sub-folders in algorithm by default.

The algorithm scripts make use of Stable Baselines 3 and especially the built-in callback functionality. An evaluation callback will assess the training performance every 10,000 steps for 10 episodes and save a checkpoint of the model if a new best mean reward has been achieved. In the script's evaluation process, the model at the end of training and the best model are used simultaneously.

Since RL algorithms are generally prone to be sample-inefficient and training can take a long time, especially on large problem sets or if parallelisation is not possible, we included the StopTrainingOnNoModelImprovement callback which observes the results of the evaluation callback and will abort the training run if no new best model could be found within a predefined time frame (default: three consecutive evaluations).

4.7.2.1 Experiment Input Options Overview

Gym supports passing optional arguments to the environment. We make use of this opportunity to provide the FLP researcher with more flexibility and less need to access the environments' code for changes. These arguments are technically optional, yet some of them are critical for the behaviour of the environment, and failing to pass them may result in errors being thrown or otherwise unexpected behaviour. Figure 4.6 summarises the implementation details from Sect. 4.6.3.

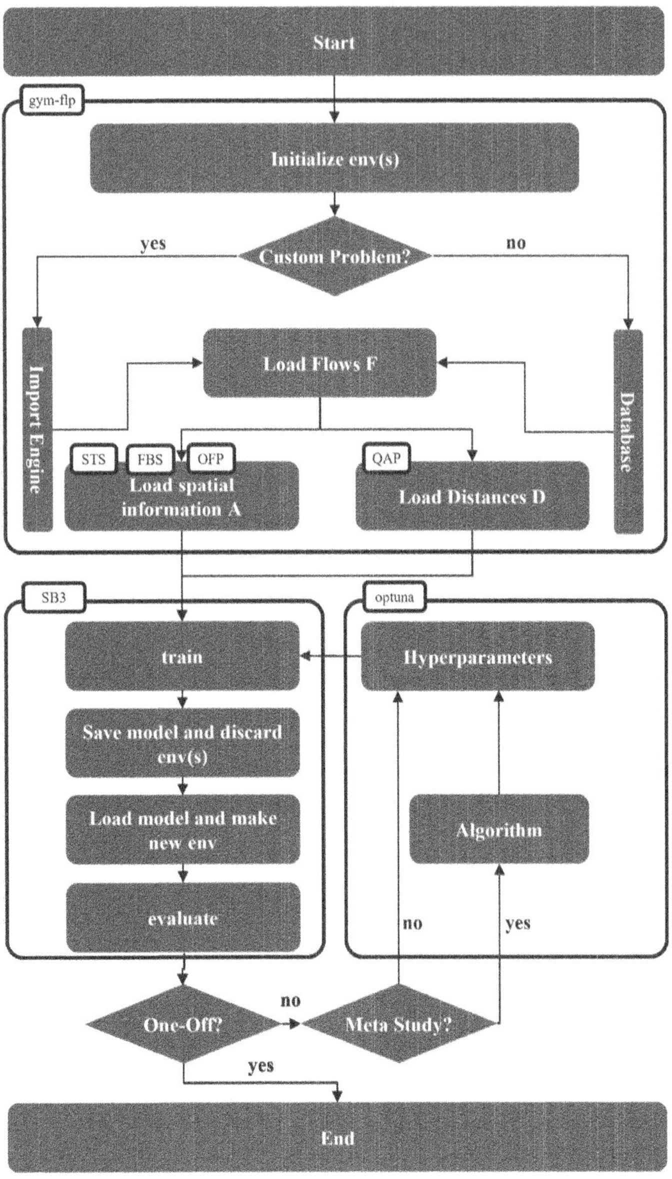

Fig. 4.5 Basic gym-flp workflow with decision gates

Scope	Characteristics	Options					
Environment	Problem representation	QAP	OFLP	FBS	STS		
	Observation Space	1-D	2-D				
	Action Space	Discrete	Multi-Discrete	Box	Box – concurrent displacement		
	Distance	Rectilinear	Euclidean	Squared Euclidean			
	Problem Instance	Literature	Custom				
	Random start state	Yes	No				
	Step size	Int: size of movement increments, applicable to OFLP only, defaults to 1					
RL Algorithm	Algorithm	PPO	DQN	A2C	SAC	TD3	DDPG
	Number of parallel workers	Int: defaults to 1					
	Training Steps	Int: defaults to 100,000					
	Hyper-parameters	Vary according to choice of algorithm					

Fig. 4.6 Morphological Box of experimentation options supported as of gym-flp 0.2.0

For training RL algorithms, gym-flp provides executable scripts under the package directory *algorithms* that can be executed via IDE or using a command line interface such as a regular terminal in Windows. The available parameters, along with their permissible values, are shown below. The default value is highlighted in boldface.

- 'mode' (*human* or rgb_array): controls whether observations are output as 1D or 2D arrays.
- 'instance' (instance name as string): the instance to be used. If no instance is passed or if it is misspelt, the environment will prompt for new input and provide an environment-specific list of available problem sets.
- 'env' (qap-v0, fbs-v0, sts-v0, or ofp-v0): the environment id to train in
- 'distance' (*rectilinear*, *Euclidean* or *squared-Euclidean*): the distance metric to be used for MHC calculations
- 'step_size': controls displacement length for discrete steps in ofp-v0. Without effect in other environments. Defaults to 1.
- 'box': if passed, will create an actions space of type Box. If omitted, a Discrete space is created instead.
- 'multi': if passed, will create a variant of the action space defined by the argument 'box' that supports simultaneous displacements.
- 'train_steps': integer value for the number of steps to take in the environment. Defaults to 100.000.
- 'num_workers': integer value for the number of parallel environments (where the algorithm provides it). Defaults to 1.
- 'algo': Name of the algorithm used for logging. If omitted, the program will use the script name instead
- 'randomize': If set, reset() will assign facilities to random locations. If omitted, machines will be distributed in a reproducible manner

4.7.2.2 Using Problem Sets from Literature

To perform studies that compare the performance of RL approaches to that of solutions presented in contemporary literature, we provide the flow (and distance or area) information for problems commonly used in literature. We implemented 138 QAPs available from the QAPLIB (Burkard et al., 1997) and a plethora of continuous problems, which were taken from the compilation of (La Scalia et al., 2019).

It is important to note that all continuous problems are available for the respective implementations as explained in Sect. 4.6.3, but their suitability may vary. For example, problem sets where the available plant space matches the sum of facility areas may not be suitable for the greenfield scenario, as the environment may fail to find a collision-free layout. On the other hand, instances where area requirements greatly undershoot plant space availability may result in empty spaces in the RGB image representations, which can hinder computation efficiency when using image recognition policies.

For all problem sets, we deliberately chose not to include currently known lowest bounds or optimal solutions from academic literature as we consider these a 'moving target' with no true added value for this package. The reason is that we intend to also solve real industry problems without any solution known a priori. However, we note and explicitly encourage any interested researcher to adapt the open-source package by, for instance, implementing an episodic approach that terminates upon reaching the state representing the best-known layout.

The available problem sets are tabulated in the readme.md of the package.

4.7.2.3 Working with Custom Problems

New FLP instances can be loaded at run-time by providing flows and distances or dimensions (depending on which type of problem representation is chosen). The gym-flp import engine accepts text files of the types.json,.txt and.prn.

When passing the value 'custom' to the argument 'instance' upon starting the training, the user will be prompted to provide the desired problem size and the file to read. This happens twice during training (training and evaluation environment) and during testing (final model and best model environment). The second input allows users to provide an evaluation instance that is different from the training instance to test the agent's performance on new information.

In the background, the EnvBuilder processor will attempt to make sure the input file is not ill-structured. The input engine expects the flow information to come first. Ideally, text files are supplied as numeric values only with connections, e.g. separated by a white space, given as one line per machine, separated by a line feed. JSON files, on the other hand, are more complicated to engineer, but easier to parse by the engine. A*.json* should be structured as per the listing below (Fig. 4.7). Alternatively, the connections j for each i, i.e. the columns of the matrices, can be given as a list following the key 'i'. The input engine will further attempt to deduce all necessary information required for the gym-flp environment. That is, if only area data are provided with the input file, the engine will make facilities square or rectangular with integer side lengths. If no plant dimensions are supplied, the engine will return an area that is twice the size of what is occupied by the specified facilities. Furthermore, the program will raise an exception and terminate if it detects distance information in inputs supplied to a continuous environment or spatial information fed into a discrete environment, respectively.

For further details on structuring custom input files, readers are referred to the gym-flp repository, where downloadable examples are provided.

```
{
 'f':
  {
   '0':
   {'0': f_{0,0},
    ...
    'n': f_{0,n}}
   'n':{'0': f_{n,0},
    ...
    'n': f_{n,n}}
  }
}
```

Fig. 4.7 Illustration of a.json structure for submitting a custom FLP to gym-flp

4.7.2.4 Example 1: One-Off Training

This example represents an atomic version of algorithm training with a single training run followed by one evaluation run. We train once with 100,000 and once with 1,000,000 steps to compare the effect of increasing the training budget. To train using PPO, we invoke the respective scripts from a terminal as follows:

Python train.py --algo ppo --distance Euclidean

and

python train.py --algo ppo --distance Euclidean --train_steps 1000000

The results of 100,000 training steps can be found in Fig. 4.8a. One can observe a steady improvement of MHC for about 80 *evaluation* steps. Afterwards, it is likely that the agent had not yet processed a sufficient amount of training observations, leading it to propose a series of actions that do not yield improvements and eventually run into a termination criterion. With an increased training budget (Fig. 4.8b), the agent achieves even lower MHC values. In addition to this, we see a plateau starting at around 100 evaluation steps, after which the evaluation stops due to not improving for five consecutive steps.

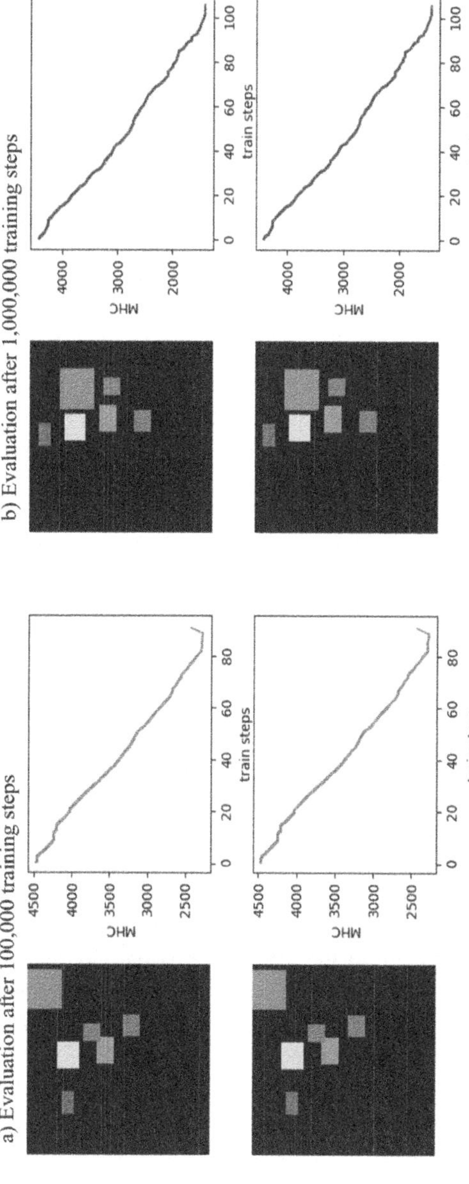

Fig. 4.8 Results of an evaluation run in ofp-v0 for 100,000 and 1,000,000 training steps in the instance P6

4.7.2.5 Example 2: Tuning Hyper-parameters with Optuna

The gym-flp training script leverages Optuna (Akiba et al., 2019). Optuna can, for instance, be used to optimise hyper-parameters of agents before submitting them to longer trials.

To do so, any parameter present in the algorithm configuration file can be passed to the training script with two values which will be interpreted as lower and upper bound in that order. The program runs 20 iterations by default, other values can be passed with tune_runs. The optimisation procedure is set up to use the relative improvement of MHC (MHC at the start—MHC at the finish) per each trial and attempts to maximise this value. The listing below shows an example for optimising the PPO hyper-parameter learning_rate.

Python train.py --algo ppo --instance P12 --learning_rate 1e6 1e2

It should be noted that RL training, especially with larger problem sets, tends to require many training steps before convergence sets in (e.g. the common instance P6 shows converging behaviour after around 1 million training steps). While it is possible to pass optimisation bounds for all hyper-parameters simultaneously to mimic a full-factorial design of experiment (DoE), it is highly unlikely that this approach will yield any meaningful results in a satisfactory time frame.

4.7.2.6 Example 3: Meta-Study

Optuna is further useful to conduct comprehensive studies to examine the effects of using different algorithms, or as originally intended with gym-flp, changes to environment design. The listing below demonstrates an experiment that performs one training run for instance P12 with 1,000,000 training steps for the three algorithms PPO, DQN and A2C (in the given order). The results of the evaluation runs for all three algorithms can be seen in Fig. 4.9.

Python train.py --instance P12 --train_steps 1e6 --algo ppo dqn a2c

Of course, this approach can be used in conjunction with Sect. 4.7.2.5. In this case, the Optuna trial will begin by suggesting the input training values and consecutively submitting them to the algorithms passed along as a list. If any of the hyper-parameters passed are not supported by the algorithms, the input values will not have any effect. No warnings are thrown in this case, so special care must be taken when designing experiments.

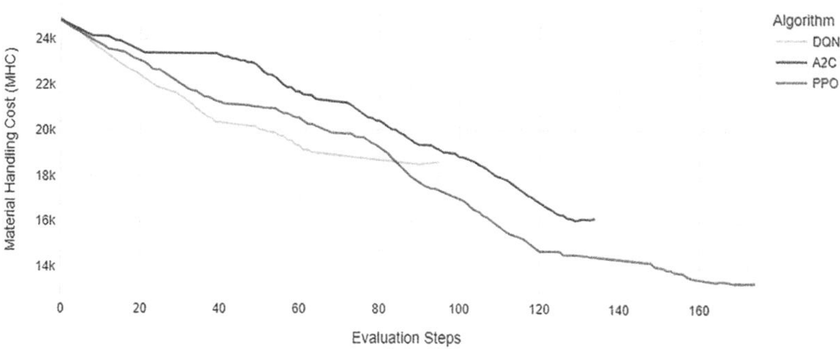

Fig. 4.9 Evaluation result of training three RL algorithms on the instance P12 for 1M steps each. It can be observed that PPO yields the best results given identical hyper-parameters as it achieves the lowest MHC

4.7.2.7 Example 4: Beyond the Package

Instead of using the experiment workflow that is built-in into gym-flp, users can of course write their own scripts for training. This is especially recommended when using RL frameworks other than Stable Baselines 3, such as Ray or a custom implementation. To achieve this, the environment side of gym-flp can be used stand-alone. A generic script as a starting point is given below in Algorithm 1.

In the context of operations research, it can be useful to replace steps 9 and 20 by other algorithmic approaches from the field. Taking the OFP as an example, the internal observation can be accessed at any time, too, simply by calling env.internal_state. The internal observation corresponds to the state output in human mode and is a vector of size $4 * n$ that holds the coordinates and dimensions of each facility (see Sect. 4.6.3.1). That information could be encoded differently outside gym-flp, e.g. as chromosomes to make it useable with genetic algorithms.

```
Initialise gym and gym-flp
M ← Initialise instance of RL model class
env ← instance of class GymEnv                        ▷ Use env=gym.make(...)
T ← Number of train steps (int)
s_0 ← env.reset()                                     ▷ obtain initial observation
for t = 1,T do
    Sample random action a from action space A
    Pass a to env and receive tuple return (o, r, d, i)    ▷ env.step(a)
    Pass tuple to RL algorithm (Update replay buffer, value function, network weights, ...) and update M
end for
Save RL model, delete env
Start evaluation run:
Make new env
eval_env ← instance of class GymEnv
o ← eval_env.reset()                                  ▷ obtain initial observation
M ← Load RL model
d ← False
mhc ← empty list
while d ≠ True do
    a ← predict a for o from M
    Pass a to eval_env and receive tuple return (o, r, d, i)
    Append i to mhc
end while
Plot/Interpret mhc
```

4.8 Section VII: Concluding Remarks

In this paper, we present a Python library intended to assist operations researchers in advancing the usage of Reinforcement Learning techniques for Facility Layout Problems. As RL has gained traction as a methodology for related combinatorial optimisation problems, we firmly believe that this is a direction worth exploring. As an initial stepping stone, we implemented three commonly used FLP problem definitions (QAP, FBS, and STS) plus an open-field layout problem (OFP), including the material handling cost computation. The package follows the conventions used in the OpenAI Gym framework, allowing the FLP instances to serve as future benchmark problems for RL in FLP research.

We consider this package in its current version as a starting point for other keen FLP researchers who can freely modify our environments by accessing their local copy of gym-flp.py and make amendments to, e.g., reward signals or even add whole new environments to it. We encourage interested researchers to explore the functionality of the package and to engage in, among others, the following activities:

- Filing issues where reproducible code problems arise
- Adding additional RL algorithms by providing an agent branch
- Writing further unit tests using test branches
- Introducing more FLP components by submitting a feature branch (drop-off points, row-based layouts, other reward mechanisms, etc.)

4.9 Declaration of Own Contribution

All work for the research was primarily carried out by the doctoral candidate. This encompasses, first and foremost, the generation of the software artefact, which includes conceptualising the individual Facility Layout Problem components, setting up the simulation environment, code repository maintenance, unit testing, and acceptance testing. Two student research assistants have supported with code generation and gathering data for benchmark problems. In addition, manuscript drafting was performed by the doctoral candidate. Both co-authors, Prof. Dr. Burggräf and Dr. Wagner, contributed to conceptualising the artefact design and to the internal review process.

5 Publication III: Deep Reinforcement Learning for Layout Planning—An MDP-Based Approach for the Facility Layout Problem

5.1 Abstract

Deep Reinforcement Learning (DRL) has demonstrated operational excellence in several production-related problems. This paper applies DRL to facility layout problems (FLP) using Proximal Policy Optimisation, Advantage Actor-Critic and Deep Q-Networks. We show that the proposed approach produces an improved arrangement of facilities. The contribution of this work is the proof of concept that DRL can optimise layouts with respect to material handling costs using only an image representation of the layout and a reward signal. The approach shows potential to generalise to new layouts without the need to model or train, thus significantly speeding up layout design procedures.

5.2 Section I: Introduction

The facility layout problem (FLP) entails finding an optimal arrangement of functional units (e.g. departments or machines) of industrial production systems. Changing dynamics of facility layout planning demands novel resolution approaches that eliminate the need for intricate re-modelling of problems and speed up the FLP planning process.

Heinbach, B.; Burggräf, P.; and Wagner, J. (2023): "Deep reinforcement learning for layout planning—An MDP-based approach for the facility layout problem". In *Manufacturing Letters* 38, pp. 40–43. https://doi.org/10.1016/j.mfglet.2023.09.007.

B. Heinbach, *Reinforcement Learning-Based Planning of Factory Layouts*, Findings from Production Management Research ,
https://doi.org/10.1007/978-3-658-51554-6_5

FLPs have predominantly been addressed with a variety of resolution approaches, e.g. (Burggräf, Wagner, Heinbach, 2021; Drira et al., 2007; Hosseini-Nasab et al., 2018; Pérez-Gosende et al., 2021), but according to recent reviews, systems using machine learning (ML) have not been used to a large extent despite their promising capabilities (Burggräf, Wagner, Heinbach, 2021; Pérez-Gosende et al., 2021). Current research has shown reinforcement learning (RL), a subset of ML, to be useful in neighbouring manufacturing fields of study, e.g. scheduling (Lee and Lee, 2022; Stricker et al., 2018; Waschneck et al., 2018), preventive maintenance (Su et al., 2022), or energy management (Weigold et al., 2021; Zhu et al., 2022).

In contrast, a recent study reports that RL has been sparsely used in FLP research (Burggräf, Wagner, Heinbach, 2021). A number of first proof-points have been published: Klar et al. used Double Deep Q-Networks (DQN) to solve an allocation problem with four (2021) and 20 units (Klar, Hussong et al., 2022), respectively, yielding promising results in a short amount of time. Ikeda et al. (2022) use a state-action table where units are deployed onto a grid. The authors acknowledge the tabular method to be restrictive and propose using deep RL approaches. Di and Yu (2021) employ a DQN to arrange furniture in indoor scenes, albeit limited to four to six functional patterns.

These examples indicate that RL may ascend to be a valuable addition to the FLP research and practice toolbox. However, as Li et al. (2023) highlight, a key objective for using DRL in manufacturing is to reduce manual involvement. In fact, facility layout planning is typically a highly manual process. Deep Neural Networks possess exceptional representation learning capabilities and can abstract a problem's structure from complex high-dimensional input that are impossible to map in tabular RL approaches due to a mushrooming dimensionality. While Unger and Börner (2021) and Di and Yu (2021) propose to model the FLP as a Markov Decision Process (MDP) with the state of the environment as image input to enable generalisation, most studies do not use it or only with a small number of units. This contribution adds to the current research stream by using a problem-agnostic image input for the RL using a larger problem size.

5.3 Section II: Reinforcement Learning Methodology

Adding to the aforementioned work, we explore the use of RL for optimising facility layouts based on material handling cost (MHC). The FLP is formulated as an MDP, and an RL agent is trained and evaluated in an industrial use case. Contrary to the previous works, our approach is not an allocation problem but instead formulated as an improvement problem starting with a specific layout. Fig. 5.1 shows the underlying MDP with the action and state spaces. The environment design is described more exhaustively in (Heinbach et al., 2024).

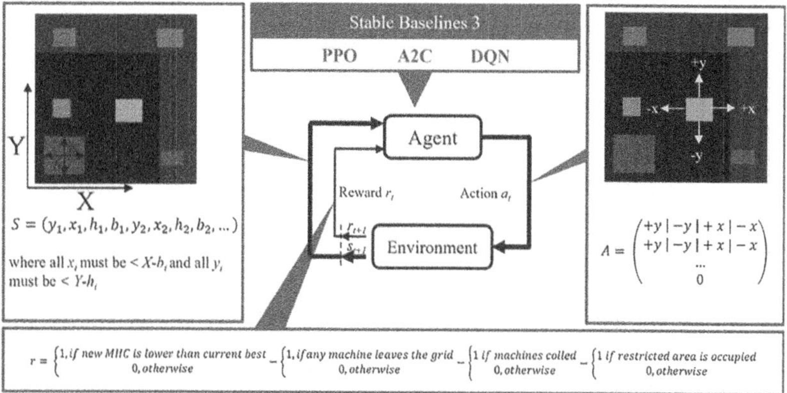

Fig. 5.1 Conceptualisation of the facility layout problem as a Markov decision process

5.3.1 State Space Representation

In line with Unger and Börner (2021) and Di and Yu (2021), we encode the state of the environment as RGB images to scale all problems to a fixed input size, allowing for easier generalisation. The proposed approach leverages a continual state space representation. Machines are positioned on a planar area of width Y and length X, and are described by their bottom-left corner coordinates and their widths b_i and lengths h_i. Distances between machines are calculated using their centroids. We restrict the state space to $Y - b_i$ and $X - h_i$ to avoid off-grid positioning. Machines are regarded as functional units that are sufficiently spaced to include drop-off/pick-up areas, storage areas, and pathways to be detailed in subsequent planning phases.

The flow information is embedded in the images' colour channels: we scale the flow matrix rows (sources) and columns (sinks) to the range in the green and blue channels. The red channel holds the machine indices, scaled between 0 and 255 to make machines distinguishable.

5.3.2 Action Space

To improve the facility layout, the agent must re-arrange the machines. We design the action space of the FLP for an agent to take one out of four possible actions on a machine, i.e. a displacement in any direction of the plant axes. Every action increments the x or y coordinate of the machine being acted upon by a step size of 1. Additionally, we include an idle action that allows the agent to remain in a state.

5.3.3 Reward

While RL agents work to maximise the cumulative reward, the goal in FLP is to minimise layout costs. As MHC is a function of distances and flows between facilities, two identical states across two different problems can yield different MHC, effectively hampering generalisability.

To discourage the agent from exploiting unwanted reward gaming loopholes (Unger and Börner, 2021), the reward signal compares the MHC in state s_{t+1} with the currently known lowest MHC, MHC_{best}, , in the episode. A reward of $+ 1$ is returned if is undershot, or 0 otherwise. MHC_{best} is reset at the start of every episode.

To ensure feasible layout solutions, we include a penalty with three terms. First, we penalise actions that move machines beyond the state space S. Second, we define a penalty for collisions if any machine F_i intersects the union of machines $F_i \dots F_n$ except itself. Third, we add a penalty if any machine is located in an otherwise restricted area that is overlaid as a mask on the state observation (red shading in the state space representation in Fig. 5.1).

Lastly, we instil episodic behaviour into our MDP. An episode is restarted if a) the agent moves a facility off the grid or b) if, during one episode, the best MHC found has not decreased for 20 consecutive steps, assuming a local optimum has been found.

5.4 Section III Case Study

We applied the discussed methodology to a real-world production scenario within the Smart Demonstration Factory Siegen, whose workshop includes twelve FUs. The material flows between facilities are based on the company's annual forecast of production quantities.

We trained algorithms from Stable Baselines 3 that support discrete action spaces: Proximal Policy Optimisation (PPO), Advantage Actor-Critic (A2C) and Deep Q-Networks (DQN), using the default hyper-parameters from (Mnih et al., 2013; Mnih et al., 2016; Schulman et al., 2017). To evaluate the agents' performance, we deployed the trained model onto a clean instance of the problem environment. The agent was tasked with making predictions until any of the specified terminal conditions presented in 2.3 were met.

Figure 5.2 shows the results for the mean episode reward during training (left) and the progress of MHC development over the rest run (right). The bottom row in Fig. 5.2 shows the layout at the beginning of evaluation (A) as well as the final layouts generated by all three agents (B-D). It is well observable that the agents exhibit learning progress. All agents propose actions that lead to decreasing MHC and thus to an improvement of the factory layout.

The final layouts of DQN (B) and A2C (C) present infeasible layout situations with units being positioned in restricted areas (indicated in dark red). Notably, given an equal training budget for all agents, PPO could decrease the MHC most (53.8%) and produces a feasible layout. From the flat section at the right-hand end of curve (D), one can see that the agent had found states with lower MHC. It can be supposed that these included intersections of units with forbidden areas, which the PPO agent was able to resolve.

The results indicate that the agents learn to produce an improved layout in terms of material handling costs merely by interacting with a visual representation of a plant layout in a similar fashion as a human planner would.

However, there are significant differences in the performance of individual RL models. It remains to be tested exhaustively how increasing the training budget alters performance. Furthermore, tuned hyper-parameters may as well produce different results.

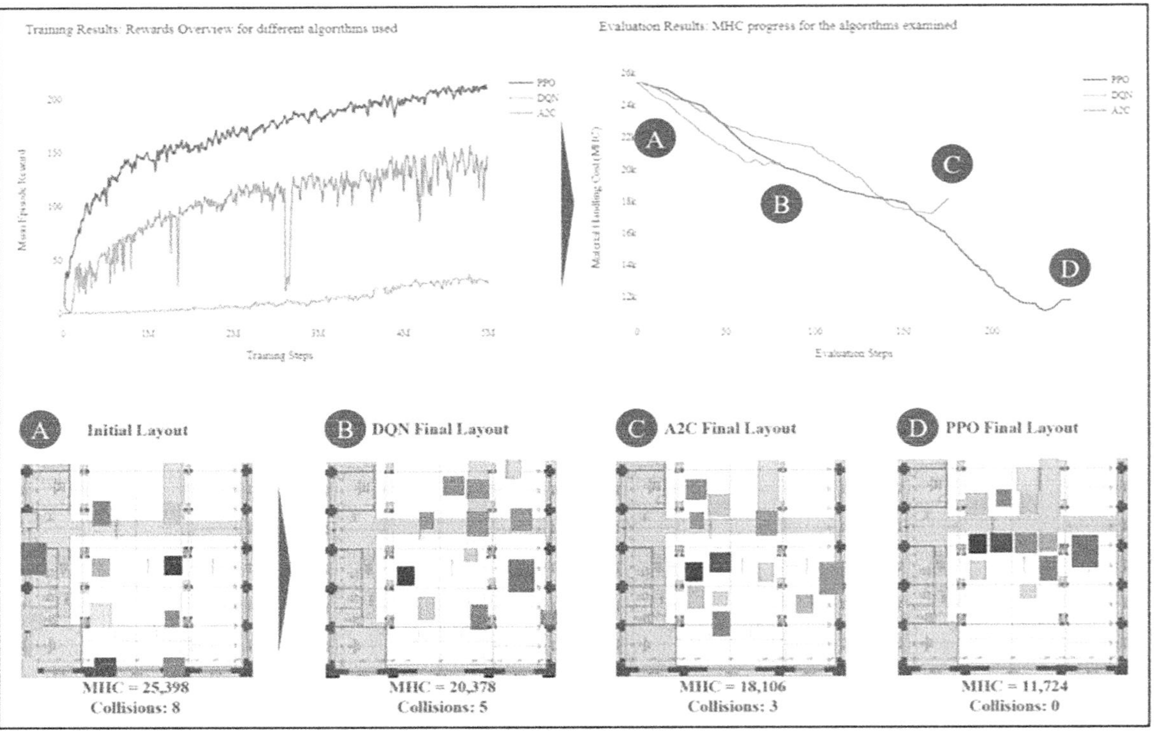

Fig. 5.2 Training and evaluation results for the algorithms PPO, DQN and A2C for 5 M training steps

5.5 Section IV Conclusion

This paper has presented a Markov Decision Process formalism to use reinforcement learning as a resolution method for the facility layout problem. We implemented the MDP in a simulation environment and trained three RL algorithms on a use case problem, and could prove that the agents can improve a layout by learning the policy of sequential machine displacements. We further demonstrated a learning success even for different criteria (reduce MHC but avoid collisions and intersections with restricted areas). This study adds to recent research on DRL for FLPs by solving a problem of larger size and by training merely from visual input. To fully leverage the representation learning capabilities in complex state spaces, future research work should explore the conditions necessary to predict policies for unseen layouts, e.g. by encoding different information in the state space.

5.6 Declaration of Competing Interest

The authors declare that they have no known competing financial interests or personal relationships that could have appeared to influence the work reported in this paper.

5.7 Declaration of Own Contribution

All work presented in the paper was primarily carried out by the doctoral candidate. This includes conceptual design, methodology development, fine-tuning of the training environment (state space and reward engineering), use case design, code generation for model performance evaluation, and conducting the experiments. Finally, manuscript drafting was performed by the doctoral candidate alone. Both co-authors, Prof. Dr. Burggräf and Dr. Wagner, contributed conceptual ideas regarding the research objective and provided in-depth manuscript review.

Publication IV: From Theory to Application: Investigating the Generalizability of Facility Layout Problems Using a Deep Reinforcement Learning Approach

6

6.1 Abstract

Generalisation is a critical requirement for reinforcement learning (RL) models applied in production planning tasks such as facility layout optimisation. This study investigates whether a single trained RL agent can generate efficient layouts across a diverse set of facility layout problems without retraining. To achieve this, we develop a scalable modelling approach using fixed-size image-based state representation and a static, masked action space, enabling consistent input-output dimensions regardless of layout size. The agent is trained using Proximal Policy Optimisation (PPO) on a subset of layout instances and evaluated on unseen problem configurations that differ in machine count and flow matrix. Material handling cost (MHC) is used as the primary performance metric, calculated as the weighted sum of Euclidean distances between machine pairs. Our results show that the trained agent significantly outperforms initial random layouts across most test cases, demonstrating meaningful generalisation. This approach offers practical benefits for scalable deployment in industrial environments where retraining for each new layout instance is often infeasible. This work establishes a foundation for adaptive, data-driven facility planning using deep reinforcement learning.

Heinbach, B.; Burggräf, P.; Steinberg, F. (2025) "From theory to application: investigating the generalizability of facility layout problems using a deep reinforcement learning approach". In *Production Engineering Research and Development.* https://doi.org/10.1007/s11740025-01352-z.

B. Heinbach, *Reinforcement Learning-Based Planning of Factory Layouts*, Findings from Production Management Research ,
https://doi.org/10.1007/978-3-658-51554-6_6

6.2 Section I: Introduction

Facility layout optimisation is critical in production engineering, directly influencing operational efficiency, cost management, and productivity (Tong, 1991). Traditional approaches for solving facility layout problems (FLPs), such as Systematic Layout Planning (SLP) (Niroomand et al., 2015) and mathematical optimisation techniques often rely on expert-driven, iterative processes that can be both time-consuming and limited in adaptability (Anjos and Vieira, 2017). As a complex combinatorial optimisation problem, FLP requires arranging machines or departments within a factory to minimise objectives such as material handling cost, travel distance, or spatial footprint. Traditionally, layout design relies on heuristics, expert knowledge, or metaheuristic algorithms. These methods often struggle to scale or adapt to changing input conditions and may not generalise across multiple instances.

In recent years, reinforcement learning (RL) has emerged as a powerful tool for sequential decision-making in complex environments, with successful applications in games, robotics, and scheduling. Previous work has introduced gym-flp, a reinforcement learning environment for FLP, and demonstrated that deep RL algorithms like Proximal Policy Optimisation (PPO) can generate efficient layouts for individual problem instances. However, these models are typically trained and evaluated on single-layout configurations, limiting their applicability in real-world settings where layout instances differ in size, flow structure, and spatial constraints.

This paper addresses a critical question for industrial deployment: Can a single RL agent generalise across a variety of layout configurations without retraining? Generalisation, in this context, refers to the agent's ability to produce feasible and efficient layouts for unseen problem instances. We propose a scalable solution that combines a fixed-size, visual representation of the layout state with a static action space and masking strategy. This formulation allows the same model architecture to be applied across multiple layout sizes and configurations, enabling inference-time adaptability. This task-specific generalisation is especially relevant for industrial use, where new layout configurations must be generated frequently under changing production conditions. By enabling rapid adaptation without retraining, the proposed method offers a scalable foundation for automated layout planning.

The contributions of this paper are threefold: (1) we present an empirical evaluation of generalisation capability in FLP using a shared visual input and masked action output across diverse benchmark instances; (2) we introduce a method that enables reusability of trained RL agents without retraining for each layout size; and (3) we demonstrate the practicality of this approach for scaling layout optimisation to dynamic or multi-product factory settings. By focusing on cross-instance generalisation, this work contributes to the broader goal of developing robust and transferable AI systems in production planning.

6.3 Section II: Background

6.3.1 Facility Layout Problem

FLPs aim to optimise the spatial arrangement of workstations, departments, machines, or equipment within a facility. This arrangement influences material handling costs (MHC), production time, worker movement, and facility utilisation. The goal is to minimise transport intensity, often represented as MHC, by considering pairwise flow relationships and distances while adhering to space, safety, and workflow constraints (Tong, 1991). The challenge in factory planning today is to design long-lasting production systems that can endure for decades while remaining flexible enough to adapt to the ever-changing market demands (Schuh et al., 2011).

FLPs are typically categorised into discrete and continuous formulations. Discrete FLPs assign components to predefined locations (grid points) and are commonly modelled as Quadratic Assignment Problems (QAPs), which seek to minimise the weighted distances between facility pairs (Lin et al., 2024). Continuous FLPs offer greater flexibility by allowing components to be placed more freely, with models such as:

- Flexible Bay Structure (FBS): Facilities are arranged in parallel bays bounded by aisles (Tong, 1991)
- Slicing Tree Structure (STS): Decomposes spaces into rectangles for optimisation (Tam, 1992)
- Open Field Layout Problem (OFP): Non-overlapping arrangements in open spaces (Niroomand et al., 2015)

In addressing FLPs, algorithms are generally categorised as construction and improvement methods (Hosseini-Nasab et al., 2018). Construction algorithms generate an initial feasible solution from scratch, often employing heuristic or rule-based approaches to establish a baseline layout that satisfies fundamental feasibility requirements, such as space limitations and workflow prerequisites. In contrast, improvement algorithms refine an existing solution through iterative enhancements, aiming to optimise the layout by reducing costs or improving efficiency. A combination of both approaches can be particularly effective: a construction algorithm provides a viable starting point, which an improvement algorithm further refines to achieve a more optimal facility layout (Chiang and Chiang, 1998).

Other approaches to FLPs include Systematic Layout Planning (SLP), which integrates both qualitative and quantitative factors in layout design but can be inadequate for highly complex problems due to its reliance on expert judgment and iterative manual adjustments (Yang et al., 2000). SLP-based approaches are suitable alternatives once the layout planning objectives are not quantifiable, which poses a major challenge for automated layout planning (Burggräf, Dannapfel et al., 2021). Where they are quantifiable, automated facility layout planning can benefit from a variety of mathematical optimisation techniques that have been developed in the past: Genetic Algorithms (GA), which are a stochastic search technique inspired by Darwinian evolution and the principle of survival of the fittest (Deb, 2011). The core component is the chromosome (or individual), representing a potential solution. GA starts with a random population that evolves over generations through genetic operators such as selection, crossover, and mutation. This iterative process optimises an objective function (fitness measure) until termination criteria are met, such as a maximum number of generations or a satisfactory solution quality (Datta et al., 2011). Simulated Annealing (SA) draws from statistical mechanics to solve combinatorial optimisation problems. It iteratively refines a solution representation by making small modifications to generate a candidate solution. If the calculated change in the objective function indicates improvement, the candidate replaces the current solution (Meller and Bozer, 1996). Particle Swarm Optimisation (PSO) is a population-based search method where particles (solutions) continuously adjust their positions in a multidimensional space. Each particle updates its position based on its own experience and that of its neighbours. Inspired by swarm behaviour (e.g., bird flocking, fish schooling), PSO is known for its simplicity and fast convergence (Paul et al., 2006). Like GA, PSO evaluates fitness and uses random techniques for optimisation, but it lacks evolutionary operators like crossover and mutation. Instead, particles update their positions using an internal velocity value and memory, following the best-performing particles. Unlike GA's two-way information sharing, PSO's approach is one-way, leading to faster convergence, though it may risk premature optimisation (Eberhart and Shi, 1998). Ant Colony Optimisation (ACO) is inspired by the foraging behaviour of real ants using pheromone trails. Initially, ants explore randomly. Once food is found, they leave a pheromone trail on their way back to the nest, attracting other ants to follow. Shorter paths accumulate more pheromones, making them more attractive, while longer paths fade due to pheromone evaporation. This self-reinforcing mechanism helps ants converge on the most efficient route. Like GA, SA, and PSO, ACO operates on a population of solutions in parallel. However, unlike the previous metaheuristics introduced, where new solutions evolve from existing ones, ACO constructs solutions from scratch in each iteration, balancing exploration and exploitation (Kulturel-Konak and Konak, 2011).

Recent interest in Machine Learning (ML), particularly Reinforcement Learning (RL), has emerged due to the success of RL in related optimisation problems such as job-shop scheduling and autonomous vehicle routing. Numerous studies have highlighted the potential of RL to address complex FLPs, reflecting a growing interest in applying data-driven methods in this field (Burggräf, Wagner, Heinbach, 2021; Klar et al., 2021; Klar, Langlotz, Aurich, 2022; Unger and Börner, 2021). (Schäfer et al., 2024) conclude that RL could dynamically adjust production systems in simulations to meet efficiency objectives in search of solutions to the durability-flexibility trade-off put forward above. Given the limitations of traditional optimisation techniques in handling dynamic and high-dimensional problem spaces, RL offers a promising alternative for FLP solutions through adaptive, experience-based decision-making.

6.3.2 Reinforcement Learning

In RL, an agent learns by interactively maximising a reward signal, aiming to develop a policy that maps states to actions to maximise the expected cumulative reward over time. Formally, RL is modelled as a Markov Decision Process (MDP) (Sutton and Barto, 2020). An MDP consists of:

- A set of states, S, that represent the possible configurations of the environment.
- A set of actions, A, that the agent can take in each state.
- A reward function, $R : S \times A \rightarrow R$, which assigns a reward to each state-action pair.
- A transition function, $T : S \times A \rightarrow S$, which specifies the next state that results from taking a given action in a given state.

At each training step, the agent selects an action based on its policy, transitions to a new state, and receives a reward (Sutton and Barto, 2020). A representation of the MDP is shown in Fig. 6.1. RL methods are commonly categorised into two

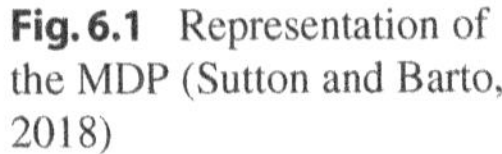
Fig. 6.1 Representation of the MDP (Sutton and Barto, 2018)

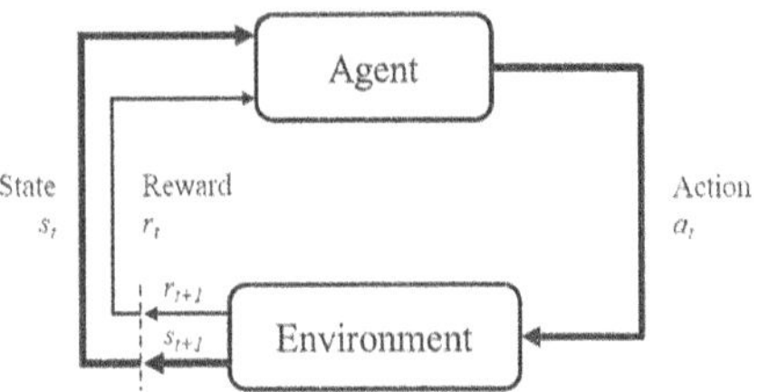

main types: model-based and model-free. Model-based RL constructs an environmental model to predict outcomes and plan actions using algorithms such as value iteration or policy iteration. While often more sample-efficient, it tends to struggle with complex environments due to the challenges of accurate modelling. Prominent representatives of model-based algorithms include AlphaZero (the successor to AlphaGoZero, which outperformed human expert Go players) (Silver, Hubert et al., 2017), world models (Ha and Schmidhuber, 2018), Imagination-Augmented Agents (Weber et al., 2017), or model-based value expansion (MVE) (Feinberg et al., 2018). Since these algorithms have little relevance to the study presented herein, we refer to the mentioned sources for in-depth insights. In contrast, model-free RL bypasses the need for an explicit model, directly learning a policy through interaction with the environment. However, this approach may require more time to converge due to increased sample complexity (Sutton and Barto, 2020).

RL methods can further be divided into value-based and policy-based approaches. Value-based methods, such as Q-learning, estimate the value of actions in different states to guide decision-making. Policy-based methods, on the other hand, directly optimise the policy itself to maximise expected rewards (Sutton and Barto, 2020).

Algorithms are also classified as either on-policy or off-policy. On-policy methods learn from data generated by the current policy, while off-policy methods, such as Q-learning, learn from data generated by different policies, which can facilitate greater exploration and improve sample efficiency (Dong et al., 2020).

A variety of algorithms have been developed to solve MDPs. The algorithms most relevant to this study are presented hereinafter.

Monte Carlo Tree Search (MCTS) is a decision-making algorithm that combines tree search with random sampling to evaluate the potential outcomes of actions in sequential decision problems. It incrementally constructs a search tree by iteratively simulating possible future states through four key steps: selection, expansion, simulation, and backpropagation. These steps balance exploration (trying less-visited actions) and exploitation (focusing on the most promising actions based on current knowledge). MCTS has been widely applied in strategic games such as Go and chess, where it has demonstrated significant success by efficiently navigating complex state spaces (Browne et al., 2012).

Deep Q-Networks (DQN) are a reinforcement learning algorithm that extends Q-learning by employing deep neural networks to approximate the Q-function, enabling the handling of high-dimensional state spaces. Rather than maintaining a Q-table, DQN uses experience replay and a target network to stabilise training, facilitating the learning of optimal policies directly from raw sensory input, such as image data. This approach has been instrumental in achieving human-level performance in complex environments, including Atari games (Mnih et al., 2015).

Proximal Policy Optimisation (PPO) is a policy-based reinforcement learning algorithm designed to balance learning efficiency and stability by optimising a clipped objective function. It constrains policy updates within a predefined trust region, preventing excessively large adjustments that could destabilise training. PPO has gained widespread adoption due to its simplicity, effectiveness, and strong performance in high-dimensional continuous action spaces (Schulman et al., 2017).

Advantage Actor-Critic (A2C) is a synchronous, policy-based reinforcement learning algorithm that combines the strengths of policy optimisation and value function approximation. It consists of two components: the actor, which updates the policy, and the critic, which estimates the value function and computes the advantage to guide policy updates. A2C is known for its efficiency and effectiveness, particularly in environments with high-dimensional state spaces, due to its ability to leverage both policy gradients and value estimation (Mnih et al., 2016).

RL has advanced significantly, with agents demonstrating exceptional performance in virtual arcade games (Mnih et al., 2015) and surpassing expert human players in board and computer games through Deep Reinforcement Learning (DRL) (Silver et al., 2016). DRL achieves this by learning directly from high-dimensional visual input (Mnih et al., 2015).

6.3.3 Reinforcement Learning in Layout Planning Problems

Burggräf et al. (2018) had highlighted that ML had become a predominant factor in production management research. Sauer and Burggräf (2024) demonstrated the importance of ML-informed decision-making in a number of production management use cases to improve operational efficiency. Their findings are relevant to FLPs, as the challenge here is also to sequentially determine the optimal level and type of human-AI collaboration.

The application of RL techniques to solve FLPs has seen increasing attention in recent years. In 2021, Burggräf, Wagner, Heinbach (2021) observed that while ML had been applied to FLPs to a limited extent, RL had not yet played a significant role. Shortly thereafter, several frameworks were proposed to explore the potential of RL in layout planning contexts. Unger and Börner (2021) were the first to formulate the facility layout problem as an MDP. They suggested encoding influential factors such as material flows, structural constraints, and placed objects into separate layers of the state representation for a neural network input. Building on this work, they later introduced an algorithm for generating optimal paths for factory layouts using a graph representation constrained by wall and workstation polygons to determine exact material flow lengths (Unger et al., 2024).

Similarly, Klar, Langlotz, Aurich (2022) propose discretising the facility layout into a grid where each grid element contains structural and object placement information, serving as an input node to a neural network. Unlike Unger and Börner (2021), their initial approach does not leverage the visual feature extraction capabilities of Convolutional Neural Networks (CNNs). However, the authors later extended their method by embedding flow information within a Graph Neural Network (GNN) and layout characteristics in a CNN, subsequently concatenating both into a joint input for the neural network's hidden layer. Furthermore, they introduced a diverse set of analytical and dynamic objective criteria (Klar, Mertes et al., 2023).

Schneidewind and Galka (2023) propose an interactive RL-based factory layout planning process. Their system architecture incorporates interaction through evaluative feedback and inverse reinforcement learning while specifying requirements for a user interface to facilitate decision-making.

Likewise, Süße et al. (2023) explore the application of RL for generative layout design. In contrast to previous approaches, they represent facility information using a $4 * M$ matrix, where M denotes the number of facilities, with each entry capturing the facility's coordinates, widths, and lengths. Notably, their concept uniquely integrates sustainability criteria into the optimisation process.

Beyond the initial framework propositions, the research field of placement and allocation solutions using RL has experienced increasing scholarly attention. Significant contributions in this area have been made by Klar and colleagues. Their initial study applied an RL-based approach to arranging four functional units within a factory layout, positioning the units near a lane designated for material transport by Automated Guided Vehicles to minimise transportation time. The state space was represented as a grid where each position was labelled as free, occupied (by a wall, lane, or functional unit), or reserved. Additionally, it included information about the next functional unit to be placed, specifying its shape and identity. The action space was defined by the grid location for placing the next functional unit and its possible rotations. To solve this problem, the authors employed Double Deep Q-Networks (DDQN), a variant of RL that stabilises training by using two neural networks and incorporates experience replay to improve learning efficiency. While the algorithm demonstrated effectiveness for the specific case with four functional units, its scalability and applicability to larger, more complex layouts remain untested (Klar et al., 2021).

Subsequently, the authors expanded the problem size to 20 functional units and included transportation metrics in the state space. By applying action masking, a modified reward structure, and a more streamlined, relevant state encoding, they demonstrated that a near-optimal solution could be achieved with improved scalability and efficiency. This approach required a reasonable training effort of 13,000 episodes compared to 8,000 episodes in their previous study (Klar, Hussong et al., 2022).

In a follow-up study investigating various reward functions, the authors found that a mixed approach—combining unit-specific and holistic throughput measures—yielded the most effective layout. The performance of the DDQN agent was compared against a manually generated layout and a computed optimal solution, with the RL-based approach achieving significant improvements in throughput efficiency while delivering results comparable to the computed optimal layout. Scalability was further tested by increasing the layout complexity to 25 functional units, where the RL agent successfully optimised throughput times, indicating robustness across varying problem sizes (Klar, Glatt, Ravani, Aurich, 2023).

A subsequent performance comparison study evaluated multiple MDP configurations on the 20-unit problem, comparing the performance of Rainbow Deep Q-Network (R-DQN), Vanilla Policy Gradient (VPG), PPO, A2C, and Soft Actor-Critic (SAC) against traditional meta-heuristic approaches such as GA, SA, Tabu Search (TS), and Adaptive Large Neighbourhood Search (ALNS). Hybrid approaches (e.g., GA-TS, GA-SA, GA-ALNS) were also included. The results showed that RL-based approaches generally performed comparably to or better than the meta-heuristic methods, with Rainbow DQN achieving the best overall performance (Klar, Langlotz, Aurich, 2022).

In a subsequent study, now relying on R-DQN, the authors incorporated a GNN to model transport intensities and a CNN to encode the spatial layout representation. The approach optimised multiple objectives simultaneously, including throughput time, transportation intensity, material flow clarity, media supply, and floor-bearing capacity. Using a larger problem set consisting of 43 functional units, the RL-based solution once again outperformed meta-heuristic approaches.

Notably, the study demonstrated RL's capacity to learn transferable strategies, allowing pre-trained agents to adapt to new problems with similar structural characteristics. This capability significantly reduced computational effort while maintaining high solution quality (Klar et al., 2024).

Zhao and Duan (2024) present a DRL approach for optimising workshop facility layouts, specifically targeting dual-objective optimisation: minimising transport distances while maximising workshop area efficiency. The authors employ an actor-critic DRL framework that decomposes the layout optimisation problem into multiple sub-problems, leveraging a pointer network for solution generation. To enhance efficiency, a parameter transfer strategy is introduced, enabling the model to iteratively solve each sub-problem while transferring learned parameters between neighbouring solutions. This approach reduces computational overhead and accelerates convergence. Experimental results demonstrated that the DRL-based method produced more efficient layouts than traditional algorithms in real-time, achieving significant reductions in logistics volume and layout area.

Another approach explores the combination of meta-heuristics with reinforcement learning (RL). Lin et al. (2024) address the UA-FLP with scenarios involving up to 62 facilities, proposing a Learning-Based Simulated Annealing (LSA) algorithm that integrates RL with traditional SA. In this framework, the layout design is performed by SA, while the RL controller dynamically guides the selection of neighbourhood operators based on feedback, enhancing the balance between exploration and exploitation. The results indicated that the proposed solution outperformed state-of-the-art algorithms, achieving superior or equivalent results in 13 out of 16 benchmark instances. Furthermore, the dynamic neighbourhood selection mechanism using RL significantly improved the convergence rate.

Ikeda et al. (2022) propose employing the MCTS method, where the agent learns through trial and error to arrange rectangular units within a 7×5 block site. The action space involves selecting and placing units, while the state space is defined as the largest unoccupied rectangular area, dynamically guiding the agent's decisions. Experimental results demonstrated that RL efficiently generated near-optimal layouts. However, the authors noted that further research is needed to extend the approach to multi-floor facilities and improve its adaptability to diverse real-world conditions.

Kaven et al. (2024) present a cooperative, asynchronous multi-agent reinforcement learning (MARL) approach to address integrated facility layout planning and job scheduling in line-less mobile assembly systems (LMAS). The study considers layouts of varying scales, with up to 20 stations constrained within a 128×128 grid. The MARL framework employs PPO, where agents manage layout decisions (station placement) and scheduling tasks (job-station assignments). The action space includes selecting potential positions for mobile stations (layout planning) and assigning jobs to stations (scheduling), while the state space encodes environmental features such as grid-based station positions, station availability, and operational progress. The results demonstrated strong performance across diverse simulation conditions, with partial generalizability to unseen scenarios.

This research advances the application of DRL in FLP by employing a problem-agnostic image-based input, suggesting that future work could focus on enhanced representation learning to improve generalisation across novel layouts.

6.3.4 Discussion of the Literature

The valuable studies presented thus far suggest that the combination of RL and FLP is indeed a promising approach. However, it also becomes evident that the field remains relatively nascent.

In traditional ML, the capacity of a trained model to operate effectively under novel circumstances and with unseen data is typically assessed using metrics such as accuracy or precision for classification problems and mean squared error for regression tasks (Flach, 2019). While DRL has demonstrated robustness within tested scenarios, its generalizability to novel or unseen facility configurations has not been extensively examined on a large scale. This gap suggests a need for further research to confirm its applicability across diverse problem settings. Notable exceptions include Klar et al. (2024), who highlighted improved training efficiency when using a pre-trained RL model, and Kaven et al. (2024), who emphasised that modifications to encoder inputs significantly altering the problem representation can lead to a decline in RL algorithm performance.

Nearly all the investigated contributions rely on fixed-length episodes in which a predefined number of machines is positioned using a constructive rather than an improvement-based approach. Additionally, the predominant method for layout representation involves grid-encoded state spaces. While effective, these approaches risk requiring extensive re-modelling efforts prior to each layout planning process. As the authors of this paper argue, RL-generated layouts should serve as inputs to subsequent manual planning processes rather than as standalone solutions.

Therefore, we propose further investigation into the capability of RL to operate across diverse problem configurations. To achieve this, we

- employ a PPO algorithm due to its versatility in high-dimensional problems
- use a fixed-size visual image input (64×64 pixels) as state space for the RL agent, and
- following an improvement-type procedure starting from an initial layout as opposed to construction methods by and large dominating the RL in FLP research landscape

Our methodological considerations for these objectives are presented next, followed by the experimental results in section 4, and a discussion of the results in section 5.

6.4 Section III: Method

6.4.1 Problem Setup

This study focuses on assessing the generalisation capabilities of a reinforcement learning agent across varying facility layout problem (FLP) configurations. Unlike conventional optimisation research, the objective here is not to find globally optimal layouts, but rather to observe how well a model trained on certain layout sizes and flow structures can transfer its learned policy to novel, unseen instances. Layout optimisation serves only as a performance metric—specifically, material handling cost (MHC)—to quantify generalisation outcomes. We emphasise that our goal is to evaluate learning transferability, not to propose a new layout optimisation method per se.

This distinction is important in light of the broader RL literature. While robustness in machine learning often refers to noise tolerance or adversarial stability within a fixed task, we study a different notion of robustness: generalisation across structurally distinct tasks within the same problem class. That is, our focus lies in the model's ability to solve new FLPs with different numbers of machines and flow matrices -problems that require distinct spatial reasoning strategies. This is a domain-specific form of generalisation with real-world relevance, as layout conditions often change due to evolving production requirements.

To enable such cross-instance generalisation, we adopt a fixed-size input representation using an RGB image encoding of the layout state. This image encodes machine placements and flow intensities, making it agnostic to the absolute number of machines. In parallel, we define a static action space covering all possible positions on the layout grid. Action masking is used to enforce feasibility constraints, such as avoiding collisions and respecting one-machine-per-cell placement. Material handling cost (MHC) is used as the primary evaluation metric. It is defined as the total weighted distance between machine pairs and is calculated as:

$$MHC = \sum_{i=0}^{n} \sum_{\substack{j \\ i<j}}^{n} f_{ij} * d_{ij} \tag{6.1}$$

where f_{ij} represents the flow intensity between machines i and j, and d_{ij} is the Euclidean distance between their assigned positions in the layout. This formulation, widely used in FLP literature, serves as a proxy for transport effort. We assume unit transport cost and exclude additional factors such as aisle constraints or congestion to isolate generalisation behaviour in the learned policy.

This setup allows the trained PPO agent to maintain consistent observation and action structures across different problem sizes, supporting scalable transferability.

6.4.2 Experimental Design and Leave-One-Out Procedure

To evaluate generalisation, we adopt a leave-one-out (LOO) protocol across layout sizes. For each test case, the agent is trained on all but one layout size (e.g., 10, 12, 14, 16 machines) and evaluated on the left-out one (e.g., 18 machines). This ablation-style procedure allows us to assess how well the learned policy transfers to previously unseen layout configurations.

We evaluate the model by applying the trained policy to the left-out layout size. The model sequentially places machines into an empty grid, using the learned policy. After all placements are made, the layout is evaluated based on the material handling cost (MHC). The baseline for comparison is the initial layout, which is generated by assigning machines to random, non-overlapping grid positions without optimisation. This uniformly distributed configuration reflects a common heuristic starting point in facility layout literature, allowing for consistent and neutral performance benchmarking across instances. Improvements in MHC indicate that the policy has generalised effectively to the new task.

The study utilises the gym-flp package (Heinbach et al., 2024), a customisable training environment that provides a visual input to the RL model. Changing the flow matrix of the problem alters the encoding of the machines within the facility, thereby yielding a unique observation independent of machine positioning. Furthermore, we use the PPO algorithm in the StableBaselines3 (SB3) library (Raffin et al., 2019). PPO is a policy gradient actor-critic network, which is known to be capable of dealing with both discrete and continuous control tasks and achieves better sample efficiency and stability due to avoiding large policy updates through clipping rewards (Schulman et al., 2017). We further trust in the exploration abilities of PPO as demonstrated by La Fuente and Guerra (2024), which we believe to be necessary. To counter the sensitivity to hyperparameter settings as found in the same study, we run a hyperparameter optimisation procedure before initiating the actual generalisation study. In addition to this, in a previous study, we compared the performance of PPO, A2C, and DQN on a 12-facility use case with characteristics identical to those used in this work. I.e., the MDP state space is visually represented as RGB images, encoding machine locations and material flow data to facilitate generalisation across varied layouts. The results showed that, given an identical training budget, PPO achieved the most significant reduction in MHC and produced feasible layouts, while DQN and A2C led to infeasible configurations due to restricted area violations (Heinbach et al., 2023).

To deal with image inputs, SB3-PPO uses the "Nature CNN" implementation of (Mnih et al., 2015), see Table 6.1 for the neural network architecture.

Table 6.1 Convolution network architecture from Mnih et al. (2015) containing 3 convolution layers followed by flatten and linear layers. 1) SB3 programmatically determines the number of input channels, which differs here (3 for RGB) from the original 4

Layer	Input Shape	Output Shape	Kernel Size	Stride
Conv-1 + ReLU	3[1]	32	8	4
Conv-2 + ReLU	32	64	4	2
Conv-3 + ReLU	64	64	3	1
Padding	–	121*64	–	–
Linear	121*64	512	–	–

For this study, commonly referenced FLP benchmark problems from the literature are used: P4 (Amaral, 2006), TL5, TL6, TL7, TL8 (Nugent et al., 1968), D10 (Das, 1993), and P12 (Yang and Peters, 1998). The details of the problems are depicted in Table 6.3. To ensure simplicity and repeatability, the machines are uniformly distributed across the initial layout. The action space is constrained to support a maximum problem size of N = 62 which corresponds to the largest problem present in the continuous problem set available for training in gym-flp. This means that the agent has 249 actions at its disposal, grouped into chunks of 4, that represent the incremental movement of a machine in + x, -x, + y, or -y. The final action is an idle action, that the agent is supposed to select if no better layout can be achieved. Consistent with the findings of Kaven et al. (2024) and Klar et al. (2024), invalid action masking is applied by setting the action logits' probability to $1e^{-8}$ for all actions where $a > 4 * n + 1$, with n representing the number of machines in the current training or evaluation environment. The state space, in turn, is fixed to a 64×64 pixels input. It is constructed during run-time from a latent observation space that holds the x-y-coordinates of each machine as well as their widths and lengths. The reasoning behind this sizing is threefold: first, PPO in StableBaseline3 uses the CNN implementation of Mnih et al. (2015), which originally had to use a square input because of the GPU implementation it used. Second, it used to be a common heuristic to use hidden layer sizes and batch sizes as a power of 2 for more efficient matrix operations in certain GPU environments like CUDA. Third, the more pixels that need to be processed, the fewer frames per second can be achieved, and training will require more wall-clock time. If one has a large facility, say 60×60 with lean machines or narrow pathways, compressing this into a smaller input shape imposes the risk of losing such machines visually if their compressed width/length falls below 1 pixel. In essence, we consider 64×64 a solid $2^{\times}$ starting point that balances training speed and the need to include a large training set like P62 without compression loss.

The workflow of the generalizability study follows a three-step procedure:

1. Hyperparameter Optimisation for each instance
2. Model Training and Evaluation
3. Model Testing on a new problem

The eval procedure, as described in Algorithm 3, is executed after each complete training cycle of n steps. During this phase, the previously best-performing model is loaded, and a full episode is predicted using the stored policy until a terminal state is reached. As defined in Heinbach et al. (2024), the criteria for terminal states are:

- Off-Grid moves: a unit has been positioned outside the plant dimensions
- Collision: two or more units intersect
- Local Optimum: MHC has not improved in several consecutive steps (default value is 10)
- Self-termination: the agent decides to terminate the episode. This action is considered the ultimate goal, as the agent should have identified through learning that no better positioning can be achieved.

Algorithm 2: Training of RL Model

```
procedure Train(env, n_steps, α, γ, batch size)
    Initialise RL model
    env ← Instantiate gym-flp environment
    for j = 0 to n_steps do:
        Sample action acc. to policy: a_t ~ πϕ(a_t|s_t)
        Obtain reward r_t = R(s; a_t)
        Get new state s_{t+1} ~ P(s_{t+1}|s_t, a_t)
        Update policy π
        if j%1000 = 0 then
            Evaluate policy for 10 evaluation steps
            Update best model
        end if
    end for
    Call: eval(env, model)
    Return: best model
end procedure
```

The eval procedure returns the relative improvement of the MHC, calculated as (initial_{MHC} − MHC)/initial_{MHC}. Additionally, it outputs the cumulative rewards, the sequence of actions taken, and the reason for termination.

Hyperparameter tuning is conducted using Optuna (Akiba et al., 2019). Optuna employs a Tree-structured Parzen Estimator (TPE) to iteratively adjust hyperparameters, aiming to maximise the relative improvement of the material handling cost (MHC) returned by the eval procedure, as described in Algorithm 4.

Table 6.2 Core characteristics of the benchmark problems

Problem	Dimensions				Flow Matrix (going to j)											
	Plant Size	Facility	Length	Width	1	2	3	4	5	6	7	8	9	10	11	12
P4	20× 20	1	2	2	0	0	0	0								
		2	4	4	10	0	0	0								
		3	6	6	5	0	0	0								
		4	2	2	0	20	8	0								
TL5	20× 20	1	6	4	0	5	2	4	1							
		2	4	4	1	0	3	0	2							
		3	6	6	1	2	0	0	0							
		4	2	4	2	1	1	0	5							
		5	7	3	3	2	2	1	0							
TL6	25× 25	1	6	4	0	5	2	4	1	0						
		2	4	4	1	0	3	0	2	2						
		3	6	6	2	1	0	0	0	0						
		4	2	4	1	2	3	0	5	2						
		5	7	3	2	1	2	1	0	10						
		6	4.375	4	3	2	1	2	1	0						
TL7	30× 30	1	6	4	0	5	2	4	1	0	0					
		2	4	4	1	0	3	0	2	2	2					
		3	6	6	2	1	0	1	0	2	5					
		4	2	4	3	2	1	0	5	2	2					
		5	7	3	2	1	2	3	0	10	0					
		6	4.375	4	3	2	1	2	1	0	5					
		7	1.8	2	4	3	2	1	2	1	0					
TL8	30× 30	1	6	4	0	5	2	4	1	0	0	6				
		2	4	4	1	0	3	0	2	2	2	0				
		3	6	6	2	1	0	0	0	0	0	5				
		4	2	4	3	2	1	0	5	2	2	10				
		5	7	3	1	2	3	4	0	10	0	0				
		6	4.375	4	2	1	2	3	1	0	5	1				

(continued)

Table 6.2 (continued)

Problem	Dimensions				Flow Matrix (going to j)											
	Plant Size	Facility	Length	Width	1	2	3	4	5	6	7	8	9	10	11	12
		7	1.8	2	3	2	1	2	2	1	0	10				
		8	3.08	5	4	3	2	1	3	2	1	0				
D10	90× 90	1	10	15	0	0	0	0	0	0	0	0	0			
		2	8	12	10	0	0	0	0	0	0	0	0			
		3	12	15	35	0	0	0	0	0	0	0	0			
		4	9	9	18	10	16	0	0	0	0	0	0			
		5	6	9	25	18	0	18	0	0	0	0	0			
		6	18	25	45	0	53	25	63	0	0	0	0			
		7	16	20	11	4	0	11	30	0	0	0	0			
		8	5	8	0	11	30	8	16	40	0	0	0			
		9	11	11	31	30	61	29	0	13	17	0	0			
		10	13	19	16	0	8	19	21	14	79	11	35			
P12	60× 60	1	4	5	0	0	0	0	0	0	0	0	0	0	0	0
		2	5	7	18	0	0	0	0	0	0	0	0	0	0	0
		3	5	6	6	0	0	0	0	0	0	0	0	0	0	0
		4	4	4	12	0	0	0	0	0	0	0	0	0	0	0
		5	6	6	2	0	4	8	0	0	0	0	0	0	0	0
		6	4	5	20	0	4	0	10	0	0	0	0	0	0	0
		7	7	10	18	0	14	0	2	0	0	0	0	0	0	0
		8	5	7	10	10	0	30	0	34	0	36	0	0	0	0
		9	5	6	38	0	16	0	30	0	12	0	0	0	0	0
		10	5	5	20	0	36	0	6	14	20	0	8	0	0	0
		11	5	5	26	18	32	0	14	0	4	6	22	0	0	0
		12	4	6	26	0	38	0	24	0	28	0	12	0	6	0

Algorithm 3: Online Evaluation of RL Model

procedure *Eval(env, model)*

Reset env

$F \leftarrow F(env)$ ▷ *Test model on benchmark problem's flow matrix (LOO)*

Set $inital_{MHC} \Leftarrow mhc_t = 0$

Set done $\Leftarrow$ *False*

while *done* $\neq$ *True* **do**

Predict next action a_t: $a_t = \max_a \pi(a|s_t)$

Get new state s_{t+1}

$R \leftarrow r_t$

$MHC \leftarrow mhc_t$

end while

Return: $mhc_{\%} \leftarrow (initial_{MHC} - MHC)/initial_{MHC}$

end procedure

Algorithm 4: Hyperparameter Optimisation

procedure *hyperOpt(env, n_trials)*

for $i = 0 \; to \, n_{trials}$ **do**

$\alpha \sim [1e^{-6}, 1e^{-3}]$

$\gamma \sim [0, 1]$

$batch_{size} \sim [2^2, 2^{10}]$

$n_{steps} \sim [2^2, 2^{10}]$

$F \leftarrow F(env)$ ▷ *Fix F to benchmark problem*

Call: *train(env, n_steps, α, γ, batch_size)*

end for

Return: best α, γ, n_{steps} *and* $batch_{size}$

end procedure

Algorithm 5: Generalisation study

1: ***procedure*** *Generalisation(instance list, n_trials)*
2:     ***for each*** *inst in instance list* ***do***
3:         *Make env*
4:         *Fetch hyper-parameters α, γ, n_steps and batch size from Algorithm 4 results*
5:         $n \leftarrow n_{trials}$
6:         ***for*** $k = 0$ *to* n ***do***
7:             *Sample random flow matrix:* $F \sim \mathcal{U}(0,1)$
8:             $F^* \leftarrow F^* \cup \{F\}$ ▷ *Update list of flow matrices excluded from evaluation*
9:             ***Call****: train(env, n_steps, α, γ, batch size, n_steps)*
10:         ***end for***
11:         *Return: Results of evaluation run*
12:         *Begin testing of models on 100 unseen flow matrices*
13:         ***for*** $l = 0$ *to* 100 ***do***
14:             *Reset env*
15:             *Sample Random F:* $F \sim \mathcal{U}(0,1) \notin F^*$
16:             *Call: eval(env, model)*
17:             $MHC^* \leftarrow MHC^* \cup \{mhc_{\%,l}\}$
18:         ***end for***
19:     ***end for***
20:     *Return: Boxplot of generalisation study results* MHC^*
21: ***end procedure***

Finally, using the optimal hyperparameters obtained from Algorithm 4, Algorithm 5 conducts a comprehensive study. The training instances selected for the study are stored in a list. The training procedure will loop through that list (line 2 in Algorithm 5) and subject each instance to n_trials training steps (line 6 in Algorithm 5). The key distinction between the hyperparameter tuning and generalisation procedures is that the latter generates a new random flow matrix (line 7) before calling the *train* procedure (line 9). These matrices are stored to ensure that none of the test matrices were previously used during training, preventing data leakage (line 8). The value for n_trials is set to 100. After running 100 trials with 20,000 training steps each, the best model for each problem instance will be evaluated on 100 unseen configurations, i.e., new random flow matrices sampled from a uniform distribution (lines 13ff).

6.5 Section IV: Computational Results

This section presents the results of the experimental runs. Table 6.3 summarises the optimal hyperparameters identified by Algorithm 4. These values were applied in the subsequent training runs, except for the batch size. To prevent truncated mini-batches in the rollout buffer, the batch size was adjusted to be a factor of n_ steps. All training elements were performed on a laptop PC with an Intel Core i5–12450 H processor, 16 GB RAM, and an NVIDIA GeForce RTX 4060 graphics card using CUDA 12.4 GPU support for PyTorch.

Table 6.3 Hyperparameters for the tested problems

Hyperparameter	Problem						
	P4	TL5	TL6	TL7	TL8	D10	P12
N steps	128	256	1024	256	256	32	128
Learning Rate	8e–4	6.92e–5	3.48e–5	1.13e–5	1e–4	2.6e–4	2.02e–5
Discount Factor	0.0936	0.29	0.69	0.06	0.28	0.597	0.98
Batch Size	128	4	8	32	64	32	32
Policy	CnnPolicy						

While layout improvements are reported for transparency, we stress that our goal was not to outperform heuristic or exact methods, but to measure policy transferability. MHC serves solely as a proxy for generalisation effectiveness.

Figure 6.2 presents the results of the training runs across all problem instances, each conducted with 20,000 training steps on 100 unique flow matrices. The figure illustrates the progression of MHC reduction over the evaluation steps. The endpoint of each curve is annotated with the reason for termination of the corresponding evaluation run. It is observable that the trained model can produce layouts corresponding to what it identifies as local optima, evidenced by its decision to self-terminate, marked with the reason"success", for all problems up to a size of $n = 8$. FLPs with larger plant dimensions, such as D10 (90×90), require more evaluation steps compared to smaller layouts like TL5 (20×20). A similar trend is observed in the magnitude of MHC improvement. While all relative reductions in MHC for valid layouts fall within the range of 35% to 56%, FLPs with smaller machines in larger plants tend to exhibit greater relative reductions. This can also be observed in the final layouts produced, which can be seen in Fig. 6.3.

Nonetheless, it seems prudent to anchor the performance metric to a specific value, such as the relative MHC reduction, rather than relying solely on cumulative episode rewards, especially for hyperparameter tuning. Optimising for cumulative reward alone may incentivise the agent to maximise action sequences that yield incremental micro-improvements in MHC, potentially leading to inefficient learning dynamics.

As outlined in the methodology section, the model resulting from the training runs was tested against 100 new flow matrices, ensuring that none of these matrices had been previously used during training. Figure 6.35 displays the distribution of relative MHC improvements across evaluation runs that returned the termination reasons "success" or "unfeasible layout" for each problem set. Additionally, Table 6.4 reports both the mean relative improvement in MHC values and the ratio of layouts that successfully terminated with the condition "success."

As no valid layouts could be produced with 100×20,000 training steps, we further report on evaluation results for 50,000, 100,000, and 1,000,000 (D10 & P12) training steps per trial, respectively. As observed in Fig. 6.4, the median improvement in MHC for valid layouts consistently ranges between approx. 35% and 60%, similar to the benchmark flow matrix results. The interquartile range around the median remains relatively narrow at first and begins to spread starting from n = 8 functional units. The deviation in MHC improvement for n = 4 may be attributed to the larger size of problem P4 (80×80), meaning that the inter-unit distances are larger at the start. One can also note that while the problems of size n = 10 and n = 12 do not produce valid layouts despite a training budget increased by factors of 2.5, 5, and 10, there is still a significant MHC improvement on the last layout produced.

Since TL8 differs from TL7 merely through adding another machine, considerations of plant dimensions do not appear to be the root cause of the large negative spread of results from the median value. Instead, it appears that this problem size is the tipping point of an explorable state space within the given training budget.

It is crucial to emphasise that the trained generalised model demonstrated the ability to produce valid layouts with reduced MHC in at least 30% of the evaluation runs. This study highlights the generalizability of RL methods for the FLP across varying problem sizes. Although tested on a simplified layout representation, the results demonstrate that a model with a static input — specifically, a 64×64 image representation of a plant layout — and a fixed action space with an action mask can predict action sequences for any problem size included in the training set that leads to reduced MHC.

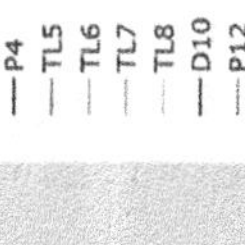

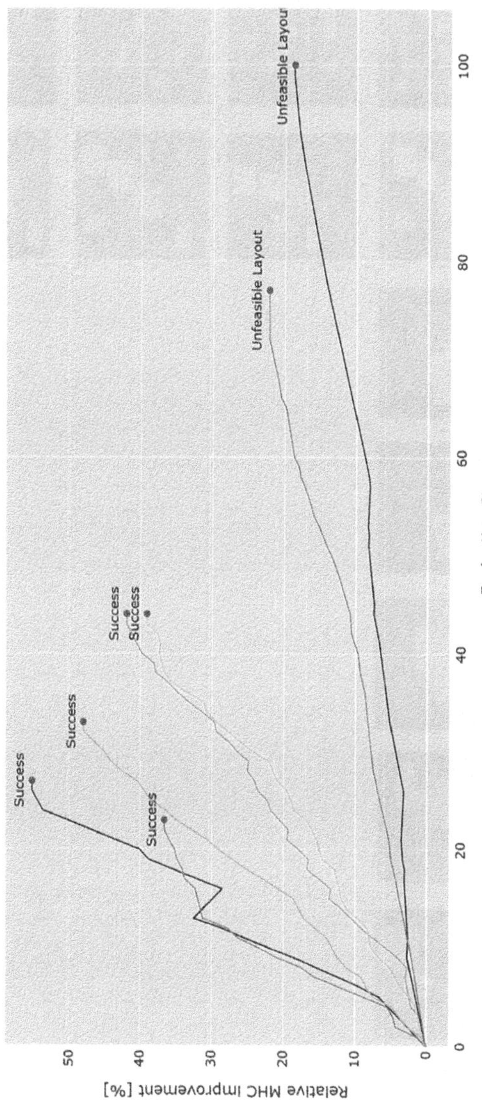

Fig. 6.2 MHC performance for each FLP and final layouts generated

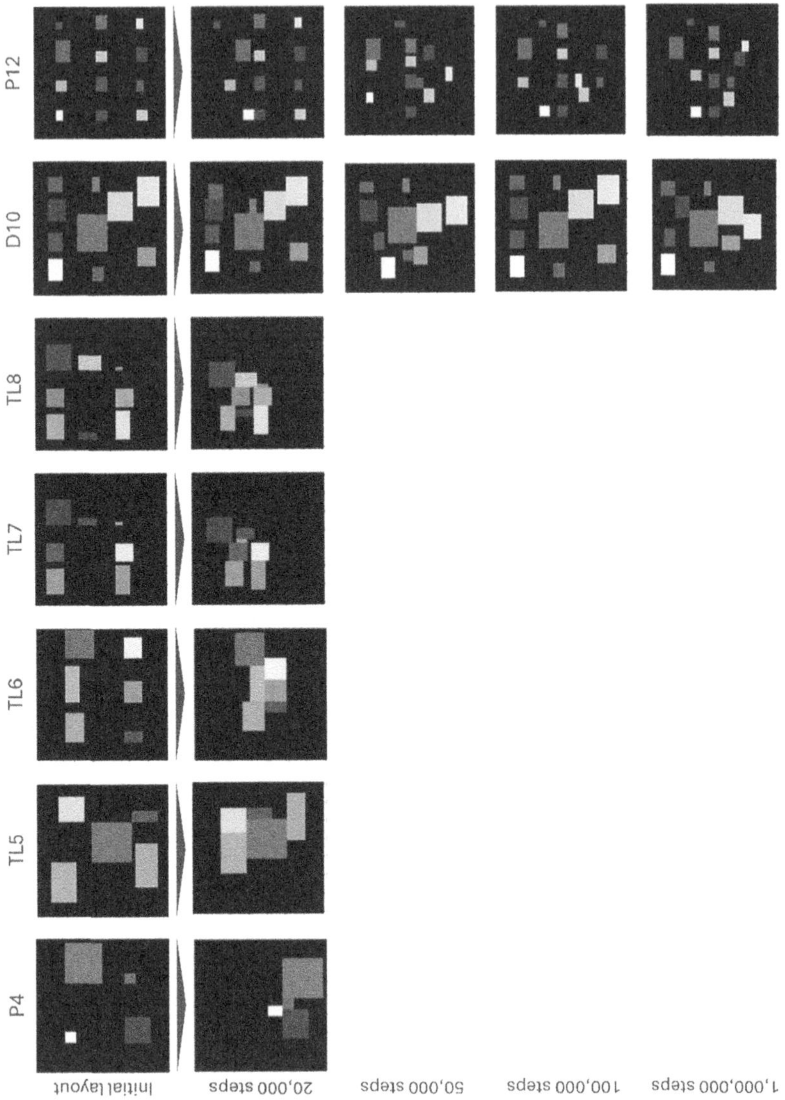

Fig. 6.3 Final layout created by the best model of the evaluation runs, per training steps

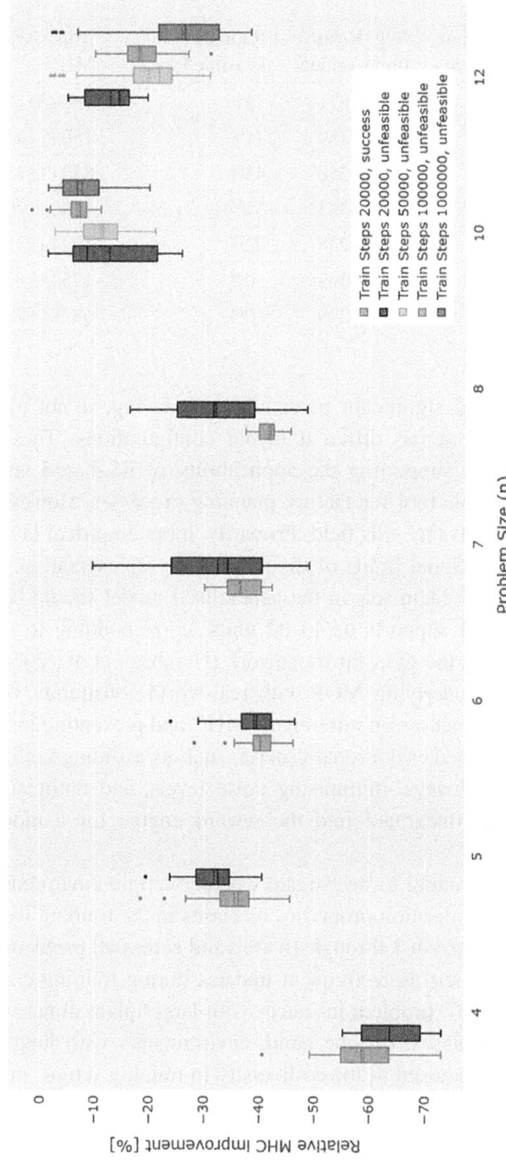

Fig. 6.4 Relative Improvements for the problem sets 4 through 12 (feasible and infeasible layouts, including suspected outliers)

Table 6.4 Achieved mean MHC improvement and layout success rate for the studied problems (20,000 steps)

Problem Size	Avg. Absolute Improvement	Avg. Relative Improvement	Ratio of Feasible Layouts	Min. (Avg.) MHC
4	–771.44	–0.592071	75%	166 (531.0)
5	–174.92	–0.355469	71%	150 (314.7)
6	–758.93	–0.401256	45%	872 (1125.2)
7	–1107.81	–0.366383	11%	1500 (1907.8)
8	–1681.73	–0.416238	22%	1924 (2355.5)
10	–24738.52	–0.127363	0%	125452 (168547.0)
12	–11541.50	–0.128299	0%	68284 (78067.8)

Such capabilities hold significant potential for reducing, if not eliminating, the need for re-training across different layout configurations. These findings provide further evidence supporting the applicability of RL-based layout planning systems as a valuable tool for factory planning processes. However, further research is required to advance this field. Primarily, more empirical knowledge is required on the computational limits of the state space representation. Secondly, the inclusion of larger problem sets in the generalised model should be pursued. The current environment supports up to 62 units, corresponding to the largest available problem within the gym-flp framework (Heinbach et al., 2024).

To better align the underlying MDP with real-world constraints, the current reward structure, which focuses on minimising MHC and preventing machine collisions, should be expanded. Additional criteria, such as avoiding collisions with structural objects or pathways, minimising noise levels, and optimising energy consumption, could be integrated into the reward engine for a more holistic optimisation approach.

Finally, it remains essential to investigate whether certain environments introduce advantageous generalisation properties or biases in the training weights. For instance, actions indexed from 1 through 16 are valid across all problem instances used, likely leading to their more frequent updates during training compared to other actions. Additionally, problem instances with larger plant dimensions could introduce bidirectional biases: on one hand, environments with longer episode lengths may have been exposed to lower diversity in training setups; on the other hand, they may contribute to a higher mean episode reward, disproportionately influencing weight updates. This suggests a potential dependence of the final results on the initial layout configuration. To mitigate this, a construction-type algorithm, which incrementally builds the layout, may be more suitable than the improvement-type approach employed in this study.

Likewise, when deploying a model, it must be assured that sufficient training has been allocated to large-scale problems. While the advantage of using an action mask is to render a model immediately usable for every problem type (within the predefined maximum number of machines to be placed, that is), it simultaneously means that virtually no generalisation can be expected for large problems if these had been underrepresented, or even omitted, in training. We purposefully choose an action space that can deal with up to 62 machines, since this is the largest available problem instance in the training set, and should, as such, be large enough to represent the majority of real-life problems.

6.6 Section V: Conclusion

The results of our study demonstrate that reinforcement learning models trained on a limited subset of facility layout problems can generalise to structurally distinct instances without retraining. This is a promising finding for scalable deployment in production planning, where factories differ in size, flow requirements, and space constraints. This distinguishes our approach from general ML robustness studies by focusing on transferability between layout instances, a form of generalisation with direct implications for scalable and low-latency deployment in industrial applications.

Importantly, this paper does not propose a novel optimisation algorithm, nor does it claim superiority in layout quality compared to state-of-the-art heuristics or exact solvers. Rather, layout performance serves as an observable metric to evaluate how effectively a trained agent transfers knowledge to new problem configurations. This distinction clarifies the contribution of our work relative to traditional layout optimisation studies. The observed generalisation suggests that the agent has learned abstract strategies that apply beyond the specific training instances, likely exploiting recurring spatial patterns or distributional similarities in the flow data. These findings support the hypothesis that general-purpose RL models, when appropriately structured, can serve as reusable planning modules for various problem classes.

Compared to classical robustness research in machine learning, which focuses on noise resistance and adversarial inputs within a static task, our focus lies in robustness across tasks -i.e., the ability to handle new instances of the same problem type with varying structural parameters. This form of generalisation is particularly important in industrial contexts, where new layouts must be generated frequently in response to changes in product design, order volume, or floorplan constraints.

Limitations of this work include the use of simplified flow matrices and a uniform initial layout baseline. While this baseline is widely used for benchmarking in RL-based layout research, it does not reflect the quality of established heuristic approaches. Future work should incorporate comparisons to standard heuristic baselines such as SLP, CRAFT, or clustering-based initialisations to provide deeper insight into performance differentials and practical competitiveness. Furthermore, it appears that the presented approach exhibits performance deterioration towards larger problems, similar to the results presented by Kaven et al. (2024). At the same time, large problems seem to benefit from a higher training budget, which unnecessarily increases training time. More work is needed to examine whether other problem representations can yield results in less time. Possible research paths could include the variable (discrete) step-size representation in gym-flp (i.e. the 'multi-box' argument) or exploring the possibilities given in gym-flp beyond discrete action spaces. Since continuous action spaces are implemented for the OFLP as well, investigating non-SB3 algorithmic approaches such as Dreamer (Danijar Hafner et al., 2020) may be a fruitful path to follow. Yet another proposal is a shift toward construction-based layout generation strategies, as opposed to purely improvement-based approaches. These may offer better adaptability in complex environments where outcomes are heavily influenced by the starting layouts. Additionally, initial layouts with large spaces between machines result in longer training episodes. Construction-type environments with fixed episode lengths may excel in terms of training time and prediction error.

Nonetheless, the current findings establish an empirical basis for generalisation in FLP and demonstrate how deep RL can support scalable, adaptive layout planning. By addressing the aforementioned areas, future research can further enhance the robustness and adaptability of RL-driven facility layout planning systems, ultimately contributing to more efficient, scalable, and intelligent decision-making frameworks for complex industrial environments.

6.7 Declaration of Own Contribution

The doctoral candidate primarily carried out all work presented in this publication. This includes conceptual design, methodology development, extending the training environment to become insensitive to varying action spaces (implementation of an action mask), benchmark problem selection, evaluation and visualisation code generation for model performance assessment, hyperparameter tuning, data gathering and analysis, and, lastly, manuscript drafting. Both co-authors, Prof. Dr. Burggräf and Dr. Steinberg, contributed conceptual ideas regarding the research objective and provided in-depth manuscript review.

7 Spotlight: A Brief Discussion on Generative AI vs. Reinforcement Learning for Facility Layout Planning

During the work on this doctoral thesis, a notable development has emerged that warrants recognition: the project commenced in early 2020. Since then, the public and widespread introduction of generative AI in the form of large language models (LLMs) has, to some extent, revolutionised everyday AI usage. In late 2022, ChatGPT-3.5 was introduced to the world by OpenAI, with GPT-4 to follow in March 2023, GPT-4o in May 2024, and GPT-o1, with reasoning capabilities, in September 2024.

Disselkamp et al. (2024) have investigated at what stages of factory planning it is suitable to use LLMs to perform tasks semi-automatically or automatically. Using GPT-4, they found that this tool could assist layout planners with a high degree of automation in some use cases along the factory planning process, primarily in information retrieval and provision. They further conclude that, due to its reliance on Dall-E, GPT-4 should not be used for generating visual layouts, unless provided in ASCII code. However, much like its successor models, GPT-4 can generate Python code, which the authors did not investigate.

As such, the simple prompt

> *"Write Python code to visualise a factory layout with 6 machines of arbitrary size in different RGB colours"*

Produces code that, once run in Python, will output the following Fig. 7.1, which in essence qualifies as a rough layout or ideal layout representation, albeit of no additional use or meaning to this point:

B. Heinbach, *Reinforcement Learning-Based Planning of Factory Layouts*, Findings from Production Management Research, https://doi.org/10.1007/978-3-658-51554-6_7

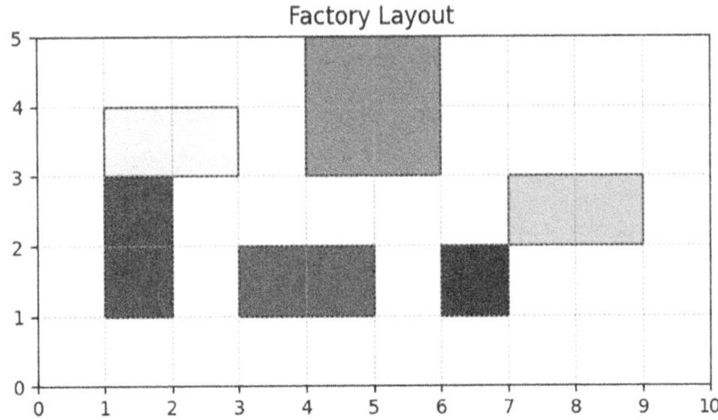

Fig. 7.1 Simple example of a factory layout generated with GPT-4

It stands to reason that a more advanced LLM is capable of being more useful in the layout planning process. Hence, using a successor to GPT-4, GPT-o1, this chapter briefly and anecdotally evaluates the performance of an LLM vs. an RL model from this thesis. Concerning the first chapter of this thesis, the underlying idea of RL-based layout planning was to learn from visual input in a way that once having learnt the structure of the problems, a trained model should be capable of providing a sequence of actions to optimise a given layout fed into it as an image (conditioned like the training examples) concerning the material handling cost. This, then, was precisely how the generalisation study in Chap. 6 was performed.

Thus, for this brief experiment, TL6 was selected as a case example; its starting layout (Fig. 7.2) was uploaded to GPT-o1, and then the following prompt was issued:

> *"The image shows a simplified factory layout of 35x35 units with 6 machines in it. The machines are depicted as rectangular objects with RGB-colored outlines. The red channel is the unit number from 1 to 6. The green channel represents outgoing flows, while the blue channel represents incoming flows. The objective is to minimise transport cost, given as sum of distance * flow between all machines. Propose an optimised factory layout."*

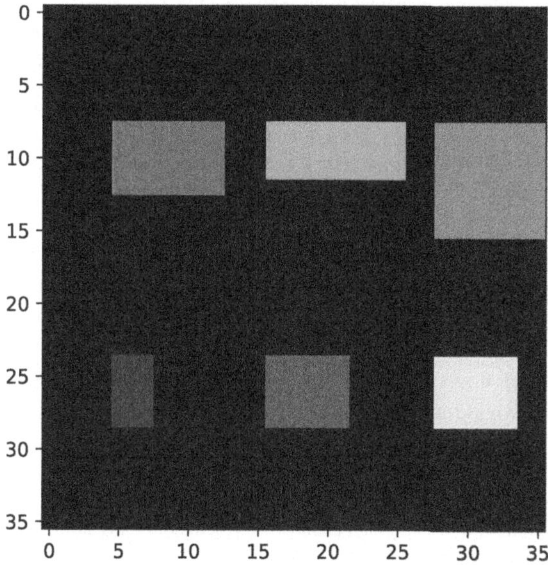

Fig. 7.2 Initial layout of TL6 as output by gym-flp that was provided to GPT-o1

After reasoning about facility layout optimisation for 17 seconds, GPT-o1 returns with a methodological example of how to improve the layout. In brief, the outline of the approach is:

1. Construct or gather the flow matrix
2. Identify the "high-flow" pairs
3. Incorporate machine dimensions
4. Generate an initial layout
5. Iterate and improve.

Interestingly, GPT-o1 itself proposes a methodology that closely resembles the triangle method of Schmigalla (Schmigalla, 1970). Thus, next, one provides GPT-o1 with the information of steps 1 through 3 (as these are given, and 4/5 are the goal from the initial prompt), and sends GPT down the journey of the example steps 4 and 5:

> *"Assume the following machine footprints: {4,4,6,4,3,4} are the heights of the machines (y dimension). {6,4,6,2,7,4.375} are the widths (x dimension), respectively. The flows are: {0,5,2,4,1,0}, {1,0,3,0,2,2}, {2,1,0,0,0,0}, {1,2,3,0,5,2}, {2,1,2,1,0,10}, {3,2,1,2,1,0}. Each block of curly braces corresponds to one machine i. The elements inside the braces are the machines j. Assume a rectangular footprint and Manhattan distance. Machine positions are given from their top-left corner. Propose a layout that minimises material handling cost. Use a solution method of your choice. Positioning should be integer values on a 35x35 grid."*

This reasoning step takes 24 seconds. The program analyses and summarises the information provided, notes relevant computations for the upcoming steps, and provides a simple Python script. Lastly, as instructed, GPT-o1 presents a layout denoted as the machine coordinates, including the explanation that machines with a large volume of flows are positioned closely. Thus, in a last step, GPT-o1 is orchestrated to provide a visualisation of the layout.

> *"Visualise the final layout. Colour the machines in RGB according to the following rules, such that they closely match the initial image provided: The red channel is the unit number from 1 to 6. The green channel represents outgoing flows, while the blue channel represents incoming flows. The machines should not have solid line borders around them. Set the background to black. Display the axis ticks in increments of 5 to represent the plant dimensions. Print the final layout cost."*

Again, after 17 seconds, GPT-o1 responds with a set of considerations, as listed below, including standalone Python code. Note that GPT-o1 had not been instructed to use Python. This may have been a result of the typical interaction of the researcher with ChatGPT. Executing the code in a Python IDE (Visual Studio Code, in this case) produces the image shown in Fig. 7.3a. Fig. 7.3b, in turn, shows the final layout produced for TL6 by the RL model of this thesis. Note that GPT-o1 slightly misunderstood the colouring instructions and plots the entire figure background in black, and the axis ticks in white. This has been corrected manually before execution.

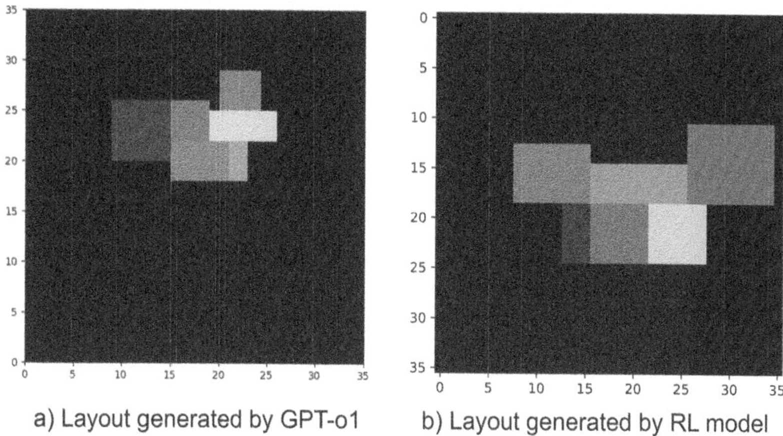

Fig. 7.3 Comparison of two layouts created by GPT-o1 and the RL model in this thesis

Table 7.1 below compares the execution times and MHC results.

Table 7.1 Summary of performance metrics for the GPT-o1 model and the RL model

Metric	GPT-o1	RL model
Final MHC	705	509
Execution time	< 1 min (reasoning) < 1 min (output) ~15 min (prompt engineering)	~300 min (training) 14.6 s (evaluation)

This illustrative chapter demonstrates two points: not only does it disprove the conclusion of Disselkamp et al. (2024) regarding the unsuitability of ChatGPT to visualise layouts. More importantly, it becomes evident that the LLM results are competitive with the RL results, both in terms of the time budget required and the results generated. Thus, it stands to reason whether extensive computing resources should be invested into one or more RL models for solving FLPs, or if a quick and lightweight LLM solution (in terms of execution time, not in processing power required in the background) is sufficiently suitable for most parts. Especially in the early stages of the layout planning process, LLM-based layout planning that can quickly ingest and process constraints in written form may pose a valuable addition to the layout planning toolbox.

8 Critical Reflection on Results and Future Perspective

The four publications within the cumulative part of this thesis focus on the layout planning of manufacturing plants. After the systematic literature review in publication I (cf. Chap. 3) to get a comprehensive overview of ML and especially RL approaches for FLPs, a software library to formalise the FLP as MDP, a proof of concept for optimising layouts with RL, and a generalisation study are presented as design artefacts in publications II through IV (cf. Chaps. 4—6). The results presented in this work demonstrate that the underlying formulation of the FLP as an MDP is valid, and that selected RL algorithms are capable of learning strategies for sequential actions that minimise a facility layout from a given initial machine positioning setup.

At the outset of this research project, as early as 2020, research and a corresponding knowledge base on the use of RL in FLP were practically non-existent. In consequence, the research question "*To what extent or subject to which limitations, can RL be a technologically viable solution to facility layout planning?*" as formulated in Sect. 1.3, can be answered as follows:

RL can be considered a technologically viable solution to facility layout planning to the extent that layout objectives can be encoded in a formal reward structure, and the problem itself can be represented as a Markov Decision Process. As demonstrated through the artefacts and results in this dissertation, especially in Chaps. 5 to 7, RL agents can autonomously learn layout policies that significantly reduce material handling costs and generalise across small to moderately sized problem instances. However, this viability is subject to several limitations: (1) performance deteriorates for large-scale layouts due to increased state and action space complexity; (2) the quality of learned policies is sensitive to the

B. Heinbach, *Reinforcement Learning-Based Planning of Factory Layouts*, Findings from Production Management Research ,
https://doi.org/10.1007/978-3-658-51554-6_8

choice of state representation, reward design, and exploration strategy; and (3) current RL methods lack integration of qualitative planning common in human expert planning. Nevertheless, within these boundaries, RL offers a scalable, adaptive, and simulation-compatible approach to layout optimisation and serves as a foundational step toward AI-driven factory planning.

Even in light of the rapid rise of generative AI and LLM-based planning tools, which will keenly assist in problem formulation and code generation, the results of this work highlight a timeless contribution: RL agents trained through environmental interaction can autonomously discover feasible and cost-efficient layouts across problem instances. This confirms that formalising FLP as an MDP and optimising spatial policies via DRL remains a valid and domain-robust paradigm, particularly where geometric constraints and simulation-grounded generalisation are required. Such constructive, grounded learning through constrained optimisation, especially for problems with no natural language formulation, remains an area where generative models are, to this end, not yet competitive. It is this experiential, artefact-based learning that makes this approach enduringly distinct and foundational regardless of technological advances.

Regarding the limitations mentioned above, the work on this topic has uncovered several key points to consider: it is necessary to scrutinise whether RL, at least under the design assumptions made in this thesis, is a suitable alternative for large-scale FLPs. To observe the RL agent's learning behaviour and performance, an improvement-type algorithm with small incremental layout changes was selected to track convergence in mean episode reward and mean episode length. A longer episode length indicates that the agent learns to avoid actions that lead to episode termination, utilising infeasible layouts. However, this approach comes with caveats: small increments not only lead to longer episodes in general but may also favour the agent in searching for the longest possible solution, i.e., the one that yields the most micro-improvements and, as such, maximises episode return. Furthermore, when approaching an optimal state, the agent must roll out the same trajectory $4 * n + 1$ times to gather the information that no further improvements are possible. Depending on the hyperparameters, the entropy loss of the policy may have decreased to a point that the agent decided to revert to exploration and lose the trajectory. A possible alternative to be investigated in the future is to change the implementation of gym-flp to a construction-based approach, which is on parts already supported by gym-flp (multi-discrete and box action spaces). However, this in turn requires reformulating the MDP to account for a scalar reward vector, which evaluates the contribution of every individual element.

The aforementioned considerations result in lower exploration within the given training budget, which may be sufficient for solving a single use case (publication III in Chap. 5) but insufficient for generalisation. In consequence, this requires the training budget to be increased in search of a generalising solution, which ultimately inflates training time. Klar et al. (2024) account for 72 hours for training their Rainbow-DQN model. The experiment on instance P12 for publication IV (Heinbach et al., 2025) took roughly the same amount of time. In personal communication, Hendrik Unger, author of Unger et al.; Unger and Börner (2024; 2021), reported that the training of his models, based on A2C, requires the use of a high-performance computing cluster. Given the indicative results of the previous chapter, the question arises as to whether the need for optimality is justified in terms of the amount of training required.

To test these assumptions, more extensive experimentation will be required. The experiments in this study have used MHC and spatial constraints (avoidance of collisions with walls and between machines) and have been limited to a maximum of twelve machines in a plant. Therefore, subsequent studies should extend problem sizes beyond $n = 12$, possibly by leveraging design changes in favour of a construction-based approach.

Further work is needed to investigate the suitability of RL methods for reducing the planning effort required by human planners in facility layout planning projects. This thesis set out with the hypothesis that RL could theoretically parallel human planning performance or, at the very least, significantly assist in the planning process by providing layouts that have already been optimised concerning quantifiable criteria, thus reducing the subsequent evaluation effort. Having successfully evaluated the artefacts designed in this work as valuable additions to the planning toolbox, the trail is blazed to broaden their usage in real-life cases. Conceivable next steps in this regard are: a) case studies with human and RL planning in tandem, with the RL Agent proposing layouts and human expert evaluating these, or b) an extension of Disselkamp et al. (2024) that analyses the interplay and planning support aptitude of a joint consideration of generative AI and RL methods.

As outlined in Sect. 2.1, real-life facility layout problems are concerned with more planning criteria than the MHC alone. However, the transfer into the practical problem domain is not within the scope of a fundamental work, as is the case with this dissertation. Thus, the upcoming research agenda for RL in FLP will need to include the addition of further optimisation objectives. To achieve this, the implementation of gym-flp can easily be adapted to incorporate additional criteria in the observation representation. Figure 8.1 below illustrates an exemplary future implementation that addresses various additional optimisation criteria. As with the current implementation from paper IV, the observation space remains fixed to a given dimension, 64×64 in this case, to allow for a problem-agnostic input. As mentioned above, larger problem sets should also be investigated (1). Drawing from the experiences of Paper III, an observation mask should be applied to model restricted areas, such as pathways and similar features, that hinder machine placements (2). Similar to (2), preferential areas (e.g., foundations, cranes) can also be modelled (3). The difference to (2) is that the element-wise AND computation rewards overlaps rather than penalising them. Equally important in real-life factory planning projects is the provision of media to workstations (electricity, compressed air, water, Ethernet, and so forth). As shown in (4), points of interest can be supplied, and the individual machine's distance to these points can be used as a reward term. Using these points as pick-up or drop-off points, and combining layers (4) and (2) can prove to be a valuable metric for machine serviceability through forklifts, etc. Lastly, ergonomical factors have gained prominence in FLPs. Presented in (5) is an example of ambient noise, where machines are spot sources of noise, and propagation and superposition have been modelled. In a similar vein, provisional areas for lighting (e.g., improved illuminance for quality assurance of assembly tasks) and temperature (e.g., avoidance of direct sunlight for storing units of sensitive electrical equipment) are also considered. This list is non-exhaustive and can be extended almost arbitrarily, as long as the corresponding reward mechanisms can be codified in the reward structure of the environment as well.

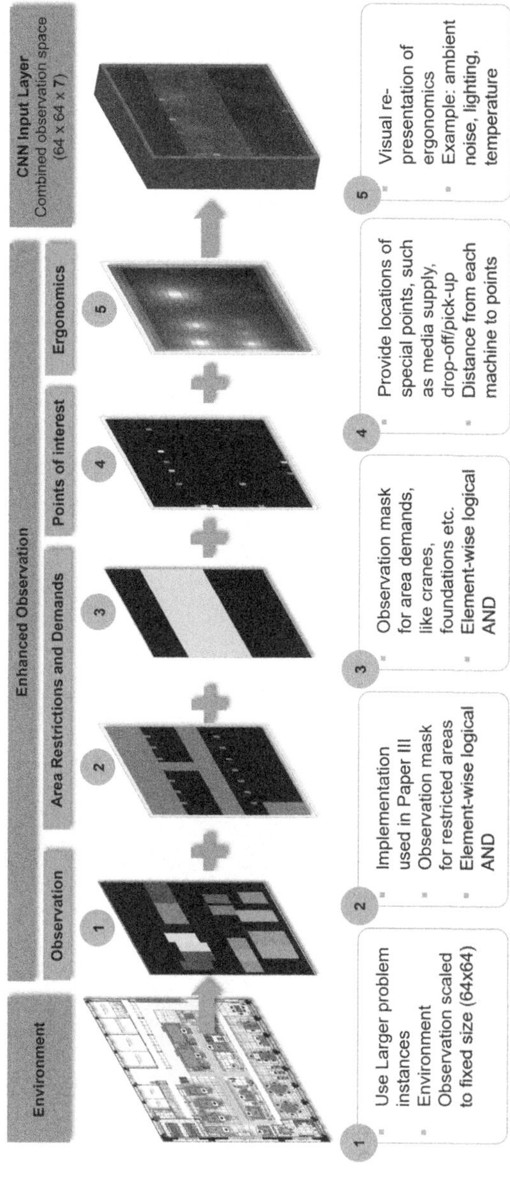

Fig. 8.1 Proposed implementations in gym-flp to address the future research scope

Summary

9

In this thesis, AI-based models were developed to design plant layouts of manufacturing facilities that are optimised concerning material handling cost and spatial constraints. During the development of these models, the following scientific results were achieved in a total of four research papers:

At the beginning, a thorough systematic literature review showed that the use of ML, and RL in particular, in otherwise extensively researched Facility Layout Problems is underrepresented. Given the variety of reviews and other works that evidence the usefulness of ML and RL approaches in neighbouring fields of production management and operations research, with similar problem structures (NP-hardness, complex search spaces), such as job-shop scheduling or path planning, this finding was remarkable. It bolstered the academic need to advance this research gap in the field.

Based on this analysis, an experimental software artefact, the Python-based library *gym-flp*, has been developed. It integrates diverse modelling approaches and design considerations regarding state and action space representations, as well as reward structures, grounded in FLP literature. It provides them as a Markov Decision Process in a simulation environment. The tool leverages OpenAI Gym, thus providing an open-source, standardised, and extensible testbed for Reinforcement Learning research in Facility Layout Problems that encompasses interfaces to off-the-shelf algorithm libraries, such as Stable Baselines or RLLib.

Subsequently, the usability of the artefacts in the problem domain has been determined in a series of experiments. First, a real-world use case confirms that state-of-the-art RL algorithms can learn to improve production layouts from

B. Heinbach, *Reinforcement Learning-Based Planning of Factory Layouts*, Findings from Production Management Research ,
https://doi.org/10.1007/978-3-658-51554-6_9

purely visual input, achieving significant cost reductions while adhering to spatial constraints. Building on these findings, a generalised training framework is introduced to address challenges of action space variability across problem sizes. The final study shows that an RL agent, trained sequentially on multiple problem instances, can generalise to unseen layouts, particularly in smaller-scale scenarios. A short discussion of the current advancements in LLMs has additionally shed light on a potential interaction between GenAI and RL, or a separation of concerns between these two.

Collectively, this work substantiates Reinforcement Learning as a viable and transferable approach for optimising dynamic and complex layout planning problems in industrial contexts. The resulting open-source software artefact empowers researchers and practitioners to efficiently apply RL in layout planning procedures. By adopting a distinct optimisation strategy—focused on iterative improvement rather than constructive generation—and by leveraging a different RL paradigm through Proximal Policy Optimisation, this research not only demonstrates practical applicability but also delivers strong systematic evidence of generalisation capabilities in this domain. In doing so, it makes a substantial contribution to the emerging field of RL in FLP and helps to close critical gaps in the existing body of knowledge.

References

Adem, A. (2023) 'A goal programming approach for the facility layout problem with ergonomic constraint', *Sigma Journal of Engineering and Natural Sciences*, vol. 41, no. 5, pp. 947–957.

Aissani, N., Beldjilali, B., Arioui, H., Merzouki, R. and Abbassi, H. A. (2008) 'On-line scheduling of Automatics and flexible Manufacturing System using SARSA technique', *AIP Conference Proceedings*. Annaba (Algeria), 30 June–2 July 2008, AIP, pp. 269–274.

Akiba, T., Sano, S., Yanase, T., Ohta, T. and Koyama, M. (2019) 'Optuna', *Proceedings of the 25th ACM SIGKDD International Conference on Knowledge Discovery & Data Mining*. Anchorage AK USA, August 4—August 8 2019. New York, NY, United States, Association for Computing Machinery, pp. 2623–2631.

Allen-Zhu, Z., Qu, Z., Richtarik, P. and Yuan, Y. (2016) 'Even Faster Accelerated Coordinate Descent Using Non-Uniform Sampling', *Proceedings of The 33rd International Conference on Machine Learning*. New York, New York, USA, June 19–24, 2016, pp. 1110–1119.

Alzubi, J., Nayyar, A. and Kumar, A. (2018) 'Machine Learning from Theory to Algorithms: An Overview', *Journal of Physics: Conference Series*, vol. 1142, p. 12012.

Amar, S. H. and Abouabdellah, A. (2016) 'Facility layout planning problem: Effectiveness and reliability evaluation system layout designs', *2016 International Conference on System Reliability and Science (ICSRS 2016) proceedings*. Paris, France, November 15—November 18 2016, IEEE, pp. 110–114.

Amaral, A. R. (2006) 'On the exact solution of a facility layout problem', *European Journal of Operational Research*, vol. 173, no. 2, pp. 508–518.

Anjos, M. F. and Vieira, M. V. (2017) 'Mathematical optimization approaches for facility layout problems: The state-of-the-art and future research directions', *European Journal of Operational Research*, vol. 261, no. 1, pp. 1–16.

B. Heinbach, *Reinforcement Learning-Based Planning of Factory Layouts*, Findings from Production Management Research ,
https://doi.org/10.1007/978-3-658-51554-6

Anjos, M. F. and Vieira, M. V. (2021) *Facility Layout: Mathematical Optimization Techniques and Engineering Applications*, Cham, Springer International Publishing.

Armour, G. C. and Buffa, E. S. (1963) 'A Heuristic Algorithm and Simulation Approach to Relative Location of Facilities', *Management Science*, vol. 9, no. 2, pp. 294–309.

Ateme-Nguema, B. and Dao, T.-M. (2009) 'Quantized Hopfield networks and tabu search for manufacturing cell formation problems', *International Journal of Production Economics*, vol. 121, no. 1, pp. 88–98.

Azimi, P. and Soofi, P. (2017) 'An ANN-based optimization model for facility layout problem using simulation technique', *Scientia Iranica*, vol. 24, no. 1, pp. 364–377.

Bahrpeyma, F. and Reichelt, D. (2022) 'A review of the applications of multi-agent reinforcement learning in smart factories', *Frontiers in robotics and AI*, vol. 9, p. 1027340.

Baker, B., Kanitscheider, I., Markov, T., Wu, Y., Powell, G., McGrew, B. and Mordatch, I. (2019) *Emergent Tool Use From Multi-Agent Autocurricula* [Online]. Available at http://arxiv.org/pdf/1909.07528v2 (Accessed 26 June 2025).

Balakrishnan, J. and Cheng, C. H. (2000) 'Genetic search and the dynamic layout problem', *Computers & Operations Research*, vol. 27, no. 6, pp. 587–593.

Balakrishnan, J., Cheng, C.-H. and Wong, K.-F. (2003) 'FACOPT: a user friendly FACility layout OPTimization system', *Computers & Operations Research*, vol. 30, no. 11, pp. 1625–1641.

Balakrishnan, J., Jacobs, F. and Venkataramanan, M. A. (1992) 'Solutions for the constrained dynamic facility layout problem', *European Journal of Operational Research*, vol. 57, no. 2, pp. 280–286.

Barat, S., Khadlikar, H., Meisheri, H., Kulkarni, V., Baniwal, V., Kumar, P. and Gajrani, M. (2019) 'Actor Based Simulation for Closed Loop Control of Supply Chain using Reinforcement Learning', *Proceedings of the 18th International Conference on Autonomous Agents and MultiAgent Systems*. Montréal, QC, Canada, May 13–17, 2019. Richland, SC, International Foundation for Autonomous Agents and Multiagent Systems, pp. 1802–1804.

Bellman, R. (1957) *Dynamic programming*, Princeton, NJ, USA, Princeton Univ. Press.

Beltran-Hernandez, C. C., Petit, D., Ramirez-Alpizar, I. G., Nishi, T., Kikuchi, S., Matsubara, T. and Harada, K. (2020) 'Learning Force Control for Contact-Rich Manipulation Tasks With Rigid Position-Controlled Robots', *IEEE Robotics and Automation Letters*, vol. 5, no. 4, pp. 5709–5716.

Bengio, Y., Lodi, A. and Prouvost, A. (2021) 'Machine learning for combinatorial optimization: A methodological tour d'horizon', *European Journal of Operational Research*, vol. 290, no. 2, pp. 405–421.

Benjaafar, S. (2002) 'Modeling and Analysis of Congestion in the Design of Facility Layouts', *Management Science*, vol. 48, no. 5, pp. 679–704.

Berlec, T., Potočnik, P., Govekar, E. and Starbek, M. (2014) 'A method of production fine layout planning based on self-organising neural network clustering', *International Journal of Production Research*, vol. 52, no. 24, pp. 7209–7222.

Besbes, M., Zolghadri, M., Costa Affonso, R., Masmoudi, F. and Haddar, M. (2020) 'A methodology for solving facility layout problem considering barriers: genetic algorithm coupled with A* search', *Journal of Intelligent Manufacturing*, vol. 31, no. 3, pp. 615–640.

Boland, A., Cherry, G. and Dickson, R. (2017) *Doing a Systematic Review: A Student's Guide*, 2nd edn, Los Angeles, USA, SAGE Publications.

Bouramtane, K., Kharraja, S., Riffi, J., El Beqqali, O. and Chraibi, A. (2024) 'A comprehensive review of static and dynamic facility layout problems', *Annual Reviews in Control*, vol. 58, p. 100970.

Brey, P. (2005) 'The Epistemology and Ontology of Human-Computer Interaction', *Minds and Machines*, vol. 15, 3–4, pp. 383–398.

Brockman, G., Cheung, V., Pettersson, L., Schneider, J., Schulman, J., Tang, J. and Zaremba, W. (2016) *OpenAI Gym* [Online]. Available at http://arxiv.org/abs/1606.01540 (Accessed 22 December 2024).

Brown, N. and Sandholm, T. (2018) 'Superhuman AI for heads-up no-limit poker: Libratus beats top professionals', *Science*, vol. 359, no. 6374, pp. 418–424.

Browne, C. B., Powley, E., Whitehouse, D., Lucas, S. M., Cowling, P. I., Rohlfshagen, P., Tavener, S., Perez, D., Samothrakis, S. and Colton, S. (2012) 'A Survey of Monte Carlo Tree Search Methods', *IEEE Transactions on Computational Intelligence and AI in Games*, vol. 4, no. 1, pp. 1–43.

Bryman, A. (2012) *Social research methods*, 4th edn, Oxford, Oxford Univ. Press.

Bryman, A. and Bell, E. (2019) *Social research methods*, 5th edn, New York, NY, Oxford University Press.

Burggräf, P., Adlon, T., Hahn, V. and Schulz-Isenbeck, T. (2021) 'Fields of action towards automated facility layout design and optimization in factory planning—A systematic literature review', *CIRP Journal of Manufacturing Science and Technology*, vol. 35, pp. 864–871.

Burggräf, P., Dannapfel, M., Hahn, V. and Preutenborbeck, M. (2021) 'Uncovering the human evaluation of changeability for automated factory layout planning: an expert survey', *Production Engineering*, vol. 15, 3–4, pp. 285–298.

Burggräf, P., Wagner, J. and Heinbach, B. (2021) 'Bibliometric Study on the Use of Machine Learning as Resolution Technique for Facility Layout Problems', *IEEE Access*, vol. 9, pp. 22569–22586.

Burggräf, P., Wagner, J. and Koke, B. (2018) 'Artificial intelligence in production management: A review of the current state of affairs and research trends in academia', *International Conference on Information Management and Processing (ICIMP)*, January 12–14, 2018. London, IEEE, pp. 82–88.

Burggräf, P., Wagner, J., Koke, B. and Bamberg, M. (2020) 'Performance assessment methodology for AI-supported decision-making in production management', *Procedia CIRP*, vol. 93, pp. 891–896.

Burkard, R. E., Karisch, S. E. and Rendl, F. (1997) 'QAPLIB—a quadratic assignment problem library', *Journal of Global Optimization*, vol. 10, no. 4, pp. 391–403.

Cadavid, J. P. U., Lamouri, S., Grabot, B. and Fortin, A. (2019) 'Machine Learning in Production Planning and Control: A Review of Empirical Literature', *IFAC-PapersOnLine*, vol. 52, no. 13, pp. 385–390.

Caruana, R. and Niculescu-Mizil, A. (2006) 'An empirical comparison of supervised learning algorithms', *Proceedings of the 23rd international conference on Machine learning*. New York, NY, ACM, pp. 161–168.

Casas, N. (2017) *Deep Deterministic Policy Gradient for Urban Traffic Light Control* [Online]. Available at http://arxiv.org/pdf/1703.09035.

Chemim, L. S., Sotsek, N. C. and Kleina, M. (2021) 'Layout optimization methods and tools: A systematic literature review', *GEPROS. Gestão da Produção, Operações e Sistemas*, vol. 16, no. 4, pp. 59–81.

Chen, F. F. and Sagi, S. R. (1995) 'Concurrent design of manufacturing cell and control functions: A neural network approach', *The International Journal of Advanced Manufacturing Technology*, vol. 10, no. 2, pp. 118–130.

Chia, R. (2002) 'The Production of Management Knowledge: Philosophical Underpinnings of Research Design', in Partington, D. (ed) *Essential Skills for Management Research,* Thousand Oaks, SAGE Publications.

Chiang, W.-C. and Chiang, C. (1998) 'Intelligent local search strategies for solving facility layout problems with the quadratic assignment problem formulation', *European Journal of Operational Research*, vol. 106, 2–3, pp. 457–488.

Chin-Sheng, C. and Kengskool, K. (1990) 'An AutoCAD-Based Expert System For Plant Layout', *Computers & Industrial Engineering*, vol. 19, 1–4, p. 299.

Chraibi, A., Kharraja, S., Osman, I. H. and Elbeqqali, O. (2019) 'A Multi-Agents System for Solving Facility Layout Problem: Application to Operating Theater', *Journal of Intelligent Systems*, vol. 28, no. 4, pp. 601–619.

Chung, Y. K. (1999a) 'A neuro-based expert system for facility layout construction', *Journal of Intelligent Manufacturing*, vol. 10, no. 5, pp. 359–385.

Chung, Y. K. (1999b) 'Application of a cascade BAM neural expert system to conceptual design for facility layout', *Computers & Mathematics with Applications*, vol. 37, no. 1, pp. 95–110.

Cioffi, R., Travaglioni, M., Piscitelli, G., Petrillo, A. and Felice, F. de (2020) 'Artificial Intelligence and Machine Learning Applications in Smart Production: Progress, Trends, and Directions', *Sustainability*, vol. 12, no. 2, p. 492.

Coraci, D., Brandi, S., Piscitelli, M. S. and Capozzoli, A. (2021) 'Online Implementation of a Soft Actor-Critic Agent to Enhance Indoor Temperature Control and Energy Efficiency in Buildings', *Energies*, vol. 14, no. 4, pp. 1–26.

Danijar Hafner, Timothy P. Lillicrap, Jimmy Ba and Mohammad Norouzi (2020) 'Dream to Control: Learning Behaviors by Latent Imagination', *8th International Conference on Learning Representations, ICLR 2020, Addis Ababa, Ethiopia, April 26–30, 2020,* OpenReview.net.

Das, S. K. (1993) 'A facility layout method for flexible manufacturing systems', *International Journal of Production Research*, vol. 31, no. 2, pp. 279–297.

Datta, D., Amaral, A. R. and Figueira, J. R. (2011) 'Single row facility layout problem using a permutation-based genetic algorithm', *European Journal of Operational Research,* vol. 213, no. 2, pp. 388–394.

Deb, K. (2011) 'Multi-objective Optimisation Using Evolutionary Algorithms: An Introduction', in Wang, L., Ng, A. H. C. and Deb, K. (eds) *Multi-objective Evolutionary Optimisation for Product Design and Manufacturing,* London, Springer London, pp. 3–34.

Deisenroth, M. P. (2011) 'A Survey on Policy Search for Robotics', *Foundations and Trends in Robotics*, vol. 2, 1–2, pp. 1–142.

del Real Torres, A., Andreiana, D. S., Ojeda Roldán, Á., Hernández Bustos, A. and Acevedo Galicia, L. E. (2022) 'A Review of Deep Reinforcement Learning Approaches for Smart

Manufacturing in Industry 4.0 and 5.0 Framework', *Applied Sciences*, vol. 12, no. 23, p. 12377.

Deng, L. (2014) 'Deep Learning: Methods and Applications', *FNT in Signal Processing (Foundations and Trends in Signal Processing)*, vol. 7, 3–4, pp. 197–387.

Di, X. and Yu, P. (2021) *Deep Reinforcement Learning for Producing Furniture Layout in Indoor Scenes* [Online]. Available at https://arxiv.org/pdf/2101.07462 (Accessed 21 December 2024).

Disselkamp, J.-P., Kurpick, D., Schutte, B., Hovemann, A. and Dumitrescu, R. (2024) 'Use Cases of Generative AI in Factory Planning: Potential and Challenges', *Proceedings of NordDesign 2024,* 12th–14th August 2024, The Design Society, pp. 196–205.

Dombrowski, U. and Marx, S. (2018) *KlimaIng—Planung klimagerechter Fabriken: Problembasiertes Lernen in den Ingenieurwissenschaften*, Berlin, Heidelberg, Springer Vieweg.

Dong, H., Ding, Z. and Zhang, S. (2020) *Deep Reinforcement Learning: Fundamentals, Research and Applications*, Singapore, Springer Singapore Pte. Limited.

Drira, A., Pierreval, H. and Hajri-Gabouj, S. (2007) 'Facility layout problems: A survey', *Annual Reviews in Control*, vol. 31, no. 2, pp. 255–267.

Duan, Y., Chen, X., Houthooft, R., Schulman, J. and Abbeel, P. (2016) 'Benchmarking Deep Reinforcement Learning for Continuous Control', *Proceedings of The 33rd International Conference on Machine Learning*. New York, New York, USA, June 19–24, 2016, pp. 1329–1338.

Eberhart, R. C. and Shi, Y. (1998) 'Comparison between genetic algorithms and particle swarm optimization', in Goos, G., Hartmanis, J., van Leeuwen, J., Porto, V. W., Saravanan, N., Waagen, D., Eiben, A. E. and Porto, V. W. (eds) *Evolutionary programming VII: 7th international conference, EP98, San Diego, California, USA, March 25–27, 1998 ; proceedings,* Berlin, Springer, pp. 611–616.

El-Rayes, K. and Khalafallah, A. (2005) 'Trade-off between Safety and Cost in Planning Construction Site Layouts', *Journal of Construction Engineering and Management*, vol. 131, no. 11, pp. 1186–1195.

Emami, S. and S. Nookabadi, A. (2013) 'Managing a new multi-objective model for the dynamic facility layout problem', *The International Journal of Advanced Manufacturing Technology*, vol. 68, 9–12, pp. 2215–2228.

Feinberg, V., Wan, A., Stoica, I., Jordan, M. I., Gonzalez, J. E. and Levine, S. (2018) *Model-Based Value Estimation for Efficient Model-Free Reinforcement Learning* [Online]. Available at http://arxiv.org/pdf/1803.00101.

Flach, P. (2019) 'Performance Evaluation in Machine Learning: The Good, the Bad, the Ugly, and the Way Forward', *Proceedings of the AAAI Conference on Artificial Intelligence*. Honolulu, Hawaii, USA, 27 January–1 February 2019, Association for the Advancement of Artificial Intelligence (AAAI), pp. 9808–9814.

François-Lavet, V., Henderson, P., Islam, R., Bellemare, M. G. and Pineau, J. (2018) 'An Introduction to Deep Reinforcement Learning', *FNT in Machine Learning (Foundations and Trends in Machine Learning)*, vol. 11, 3–4, pp. 219–354.

Friedrich, C., Klausnitzer, A. and Lasch, R. (2018) 'Integrated slicing tree approach for solving the facility layout problem with input and output locations based on contour distance', *European Journal of Operational Research*, vol. 270, no. 3, pp. 837–851.

Fujimoto, S., van Hoof, H. and Meger, D. (2018) 'Addressing Function Approximation Error in Actor-Critic Methods', *Proceedings of the 35th International Conference on Machine Learning*, PMLR, pp. 1587–1596.

Gabel, T. and Riedmiller, M. (2008) *Adaptive reactive job-shop scheduling with reinforcement learning agents* [Online]. Available at http://machine-learning-lab.com/_media/publications/gr07.pdf (Accessed 26 June 2025).

Gabel, T. and Riedmiller, M. (2012) 'Distributed policy search reinforcement learning for job-shop scheduling tasks', *International Journal of Production Research*, vol. 50, no. 1, pp. 41–61.

Gambardella, L. M. and Dorigo, M. (1995) 'Ant-Q: A reinforcement learning approach to the traveling salesman problem', in *Machine Learning Proceedings 1995*, Elsevier, pp. 252–260.

Garcia, A., Clavel, C., Essid, S. and d'Alche-Buc, F. (2018) 'Structured Output Learning with Abstention: Application to Accurate Opinion Prediction', *Proceedings of the 35th International Conference on Machine Learning*, PMLR, pp. 1695–1703.

Garcia-Hernandez, L., Garcia-Hernandez, J. A., Salas-Morera, L., Carmona-Muñoz, C., Alghamdi, N. S., Oliveira, J. V. de and Salcedo-Sanz, S. (2020) 'Addressing Unequal Area Facility Layout Problems with the Coral Reef Optimization algorithm with Substrate Layers', *Engineering Applications of Artificial Intelligence*, vol. 93, p. 103697.

Garcia-Hernandez, L., Salas-Morera, L., Garcia-Hernandez, J. A., Salcedo-Sanz, S. and Valente de Oliveira, J. (2019) 'Applying the coral reefs optimization algorithm for solving unequal area facility layout problems', *Expert Systems with Applications*, vol. 138, p. 112819.

García-Hernández, L., Palomo-Romero, J. M., Salas-Morera, L., Arauzo-Azofra, A. and Pierreval, H. (2015) 'A novel hybrid evolutionary approach for capturing decision maker knowledge into the unequal area facility layout problem', *Expert Systems with Applications*, vol. 42, no. 10, pp. 4697–4708.

García-Hernández, L., Pérez-Ortiz, M., Araúzo-Azofra, A., Salas-Morera, L. and Hervás-Martínez, C. (2014) 'An evolutionary neural system for incorporating expert knowledge into the UA-FLP', *Neurocomputing*, vol. 135, pp. 69–78.

Gawłowicz, P. and Zubow, A. (2018) *ns3-gym: Extending OpenAI Gym for Networking Research* [Online]. Available at http://arxiv.org/pdf/1810.03943v2 (Accessed 22 December 2024).

Gentsch, P. (2019) *Künstliche Intelligenz für Sales, Marketing und Service: Mit AI und Bots zu einem Algorithmic Business - Konzepte und Best Practices*, 2nd edn, Wiesbaden, Springer Fachmedien Wiesbaden.

Gough, D., Oliver, S. and Thomas, J. (2012) *An Introduction to systematic reviews*, Los Angeles, LA, Sage.

Gray, D. E. (2021) *Doing Research in the Real World*, SAGE Publications Ltd.

Greenhalgh, T. and Peacock, R. (2005) 'Effectiveness and efficiency of search methods in systematic reviews of complex evidence: audit of primary sources', *BMJ (Clinical research ed.)*, vol. 331, no. 7524, pp. 1064–1065.

Grundig, C.-G. (2021) *Fabrikplanung: Planungssystematik—Methoden—Anwendungen*, 7th edn, München, Hanser.

Ha, D. and Schmidhuber, J. (2018) 'Recurrent World Models Facilitate Policy Evolution', *Advances in neural information processing systems 31: 32nd Conference on Neural Information Processing Systems (NeurIPS 2018): Montréal, Canada, 3–8 December 2018*. Red Hook, NY, Curran Associates Inc.

Haarnoja, T., Zhou, A., Abbeel, P. and Levine, S. (2018) 'Soft Actor-Critic: Off-Policy Maximum Entropy Deep Reinforcement Learning with a Stochastic Actor', *Proceedings of the 35th International Conference on Machine Learning*, PMLR, pp. 1861–1870.

Hammad, A. W. A., Akbarnezhad, A. and Rey, D. (2016) 'A multi-objective mixed integer nonlinear programming model for construction site layout planning to minimise noise pollution and transport costs', *Automation in Construction*, vol. 61, pp. 73–85.

Harden, A., Weston, R. and Oakley, A. (eds) (1999) *Database of Abstracts of Reviews of Effects (DARE): Quality-assessed Reviews [Internet]*, Centre for Reviews and Dissemination (UK).

Hardy, Q. (2016) *Reasons to Believe the AI Boom Is Real* [Online]. Available at www.nytimes.com/2016/07/19/technology/reasons-to-believe-the-ai-boom-is-real.html (Accessed 6 June 2020).

Hart, C. (2004) *Doing Your Masters Dissertation*, Sage.

He, K., Zhang, X., Ren, S. and Sun, J. (2015) 'Delving Deep into Rectifiers: Surpassing Human-Level Performance on ImageNet Classification', *Proceedings of the IEEE International Conference on Computer Vision (ICCV)*, pp. 1026–1034.

Hein, D., Depeweg, S., Tokic, M., Udluft, S., Hentschel, A., Runkler, T. A. and Sterzing, V. (2017) 'A benchmark environment motivated by industrial control problems', *2017 IEEE Symposium Series on Computational Intelligence (SSCI)*. Honolulu, HI, USA, 27 November 2017–01 December 2017, IEEE, pp. 1–8.

Heinbach, B., Burggräf, P. and Steinberg, F. (2025) 'From theory to application: investigating the generalizability of facility layout problems using a deep reinforcement learning approach', *Production Engineering*.

Heinbach, B., Burggräf, P. and Wagner, J. (2023) 'Deep reinforcement learning for layout planning—An MDP-based approach for the facility layout problem', *Manufacturing Letters*, vol. 38, pp. 40–43.

Heinbach, B., Burggräf, P. and Wagner, J. (2024) 'gym-flp: A Python Package for Training Reinforcement Learning Algorithms on Facility Layout Problems', *Operations Research Forum*, vol. 5, no. 1.

Heragu, S. S. and Kakuturi (1997) 'Grouping and placement of machine cells', *IIE Transactions*, vol. 29, no. 7, pp. 561–571.

Hevner, March, Park and Ram (2004) 'Design Science in Information Systems Research', *MIS Quarterly*, vol. 28, no. 1, p. 75.

Hinton, G., Sejnowski, T. J. and Poggio, T. A. (1999) *Unsupervised Learning: Foundations of Neural Computation*, Cambridge, MIT Press.

Hosseini-Nasab, H., Fereidouni, S., Fatemi Ghomi, Seyyed Mohammad Taghi and Fakhrzad, M. B. (2018) 'Classification of facility layout problems: a review study', *The International Journal of Advanced Manufacturing Technology*, vol. 94, pp. 957–977.

Hu, B. and Yang, B. (2019) 'A particle swarm optimization algorithm for multi-row facility layout problem in semiconductor fabrication', *Journal of Ambient Intelligence and Humanized Computing*, vol. 10, no. 8, pp. 3201–3210.

Huang, W. and Zheng, H. (2018) 'Architectural Drawings Recognition and Generation through Machine Learning', *ACADIA 2018: Recalibration. On imprecisionand infidelity.* Mexico City (Mexico), 18–20 October, 2018, ACADIA.

Hubbs, C. D., Perez, H. D., Sarwar, O., Sahinidis, N. V., Grossmann, I. E. and Wassick, J. M. (2020) *OR-Gym: A Reinforcement Learning Library for Operations Research Problems* [Online], arXiv.

Hwang, C.-L. (1981) *Multiple Attribute Decision Making: Methods and Applications A State-of-the-Art Survey*, Berlin, Heidelberg, Springer.

Ikeda, H., Nakagawa, H. and Tsuchiya, T. (2022) 'Towards Automatic Facility Layout Design Using Reinforcement Learning', *Communication Papers of the of the 17th Conference on Computer Science and Intelligence Systems.* Sofia, Bulgaria, September 4–7, 2022, PTI, pp. 11–20.

Islier, A. A. (1998) 'A genetic algorithm approach for multiple criteria facility layout design', *International Journal of Production Research*, vol. 36, no. 6, pp. 1549–1569.

Jang, I. H. and Rhee, J. T. (1997) 'Generalized machine cell formation considering material flow and plant layout using modified self-organizing feature maps', *Computers & Industrial Engineering*, vol. 33, 3–4, pp. 457–460.

Johnson, P. and Duberley, J. (2000) *Understanding management research: An introduction to epistemology* [Online], SAGE Publications Ltd. Available at https://www.torrossa.com/gs/resourceproxy?an=4912423&publisher=fz7200.

Jolai, F., Tavakkoli-Moghaddam, R. and Taghipour, M. (2012) 'A multi-objective particle swarm optimisation algorithm for unequal sized dynamic facility layout problem with pickup/drop-off locations', *International Journal of Production Research*, vol. 50, no. 15, pp. 4279–4293.

Kang, Z., Catal, C. and Tekinerdogan, B. (2020) 'Machine learning applications in production lines: A systematic literature review', *Computers & Industrial Engineering*, vol. 149, p. 106773.

Kaven, L., Huke, P., Göppert, A. and Schmitt, R. H. (2024) 'Multi agent reinforcement learning for online layout planning and scheduling in flexible assembly systems', *Journal of Intelligent Manufacturing*, vol. 35, no. 8, pp. 3917–3936.

Khalil, E., Dai, H., Zhang, Y., Dilkina, B. and Le Song (2017) 'Learning Combinatorial Optimization Algorithms over Graphs', *Advances in Neural Information Processing Systems,* Curran Associates, Inc.

Kim, B., Jeong, Y. and Shin, J. G. (2020) 'Spatial arrangement using deep reinforcement learning to minimise rearrangement in ship block stockyards', *International Journal of Production Research*, vol. 58, no. 16, pp. 5062–5076.

Kim, J.-Y. and Kim, Y.-D. (1995) 'Graph theoretic heuristics for unequal-sized facility layout problems', *Omega*, vol. 23, no. 4, pp. 391–401.

Kim, Y. S., Shin, H. S. and Park, C. S. (2022) 'Model predictive lighting control for a factory building using a deep deterministic policy gradient', *Journal of Building Performance Simulation*, vol. 15, no. 2, pp. 174–193.

Klar, M., Glatt, M. and Aurich, J. C. (2021) 'An implementation of a reinforcement learning based algorithm for factory layout planning', *Manufacturing Letters*, vol. 30, pp. 1–4.

Klar, M., Glatt, M. and Aurich, J. C. (2023) 'Performance comparison of reinforcement learning and metaheuristics for factory layout planning', *CIRP Journal of Manufacturing Science and Technology*, vol. 45, pp. 10–25.

Klar, M., Glatt, M., Ravani, B. and Aurich, J. C. (2023) 'A simulation-based factory layout planning approach using reinforcement learning', *Procedia CIRP*, vol. 120, pp. 123–128.

Klar, M., Hussong, M., Ruediger-Flore, P., Yi, L., Glatt, M. and Aurich, J. C. (2022) 'Scalability investigation of Double Deep Q Learning for factory layout planning', *Procedia CIRP*, vol. 107, pp. 161–166.

Klar, M., Langlotz, P. and Aurich, J. C. (2022) 'A Framework for Automated Multiobjective Factory Layout Planning using Reinforcement Learning', *Procedia CIRP*, vol. 112, pp. 555–560.

Klar, M., Mertes, J., Glatt, M., Ravani, B. and Aurich, J. C. (2023) 'A Holistic Framework for Factory Planning Using Reinforcement Learning', in Aurich, J. C., Garth, C. and Linke, B. S. (eds) *Proceedings of the 3rd Conference on Physical Modeling for Virtual Manufacturing Systems and Processes,* Cham, Springer International Publishing and Imprint Springer, pp. 129–148.

Klar, M., Schworm, P., Wu, X., Simon, P., Glatt, M., Ravani, B. and Aurich, J. C. (2024) 'Transferable multi-objective factory layout planning using simulation-based deep reinforcement learning', *Journal of Manufacturing Systems*, vol. 74, pp. 487–511.

Kobayashi, M., Makita, T., Matsui, S., Koyama, M., Fujii, N., Hatono, I. and Ueda, K. (2001) 'Floor layout planning method based on self-organization', *Conference proceedings / 2001 IEEE International Symposium on Semiconductor Manufacturing: October 8–10, 2001, San Jose, California.* San Jose, CA, USA, 8–10 Oct. 2001. Piscataway, NJ, IEEE Operations Center, pp. 381–384.

Kobbacy, K. A. and Vadera, S. (2011) 'A survey of AI in operations management from 2005 to 2009', *Journal of Manufacturing Technology Management*, vol. 22, no. 6, pp. 706–733.

Kobbacy, K. A. H., Vadera, S. and Rasmy, M. H. (2007) 'AI and OR in Management of Operations: History and Trends', *Journal of the Operational Research Society*, vol. 58, no. 1, pp. 10–28.

Koke, B. and Moehler, R. C. (2019) 'Earned Green Value management for project management: A systematic review', *Journal of Cleaner Production*, vol. 230, pp. 180–197.

Konak, A., Kulturel-Konak, S., Norman, B. A. and Smith, A. E. (2006) 'A new mixed integer programming formulation for facility layout design using flexible bays', *Operations Research Letters*, vol. 34, no. 6, pp. 660–672.

Kondoh, S., Umeda, Y., Tomiyama, T. and Yoshikawa, H. (2000) 'Self Organization of Cellular Manufacturing Systems', *CIRP Annals*, vol. 49, no. 1, pp. 347–350.

Koopmans, T. C. and Beckmann, M. (1957) 'Assignment Problems and the Location of Economic Activities', *Econometrica*, vol. 25, no. 1, p. 53.

Kovács, G. and Kot, S. (2017) 'Facility layout redesign for efficiency improvement and cost reduction', *Journal of Applied Mathematics and Computational Mechanics*, vol. 16, no. 1, pp. 63–74.

Krarup, J. and Pruzan, P. M. (1983) 'The simple plant location problem: Survey and synthesis', *European Journal of Operational Research*, vol. 12, pp. 36–81.

Ku, M.-Y., Hu, M. H. and Wang, M.-J. (2011) 'Simulated annealing based parallel genetic algorithm for facility layout problem', *International Journal of Production Research*, vol. 49, no. 6, pp. 1801–1812.

Kuhnle, A., Kaiser, J.-P., Theiß, F., Stricker, N. and Lanza, G. (2021) 'Designing an adaptive production control system using reinforcement learning', *Journal of Intelligent Manufacturing*, vol. 32, no. 3, pp. 855–876.

Kuhnle, A., Röhrig, N. and Lanza, G. (2019) 'Autonomous order dispatching in the semiconductor industry using reinforcement learning', *Procedia CIRP*, vol. 79, pp. 391–396.

Kulturel-Konak, S. (2012) 'A linear programming embedded probabilistic tabu search for the unequal-area facility layout problem with flexible bays', *European Journal of Operational Research*, vol. 223, no. 3, pp. 614–625.

Kulturel-Konak, S. and Konak, A. (2011) 'Unequal area flexible bay facility layout using ant colony optimisation', *International Journal of Production Research*, vol. 49, no. 7, pp. 1877–1902.

Kusiak, A. and Heragu, S. (1987) 'The facility layout problem', *European Journal of Operational Research*, vol. 29, no. 3, pp. 229–251.

La Fuente, N. de and Guerra, D. A. V. (2024) *A Comparative Study of Deep Reinforcement Learning Models: DQN vs PPO vs A2C* [Online]. Available at http://arxiv.org/pdf/2407.14151 (Accessed 10 May 2025).

La Scalia, G., Micale, R. and Enea, M. (2019) 'Facility layout problem: Bibliometric and benchmarking analysis', *International Journal of Industrial Engineering Computations*, vol. 10, no. 4, pp. 453–472.

Le, Q. V. (2013) 'Building high-level features using large scale unsupervised learning', *IEEE International Conference on Acoustics, Speech and Signal Processing (ICASSP), 2013: 26–31 May 2013, Vancouver Convention Center, Vancouver, British Columbia, Canada ; proceedings*. Piscataway, NJ, IEEE.

LeCun, Y., Bengio, Y. and Hinton, G. (2015) 'Deep learning', *Nature*, vol. 521, no. 7553, pp. 436–444.

Lee, R. and Moore, J. M. (1967) 'CORELAP-COmputerized RElationship LAyout Planning', *The Journal of Industrial Engineering*, vol. 18, no. 3, pp. 195–200.

Lee, S., Kahng, H.-G., Cheong, T. and Kim, S. B. (2019) 'Iterative two-stage hybrid algorithm for the vehicle lifter location problem in semiconductor manufacturing', *Journal of Manufacturing Systems*, vol. 51, pp. 106–119.

Lee, Y. H. and Lee, S. (2022) 'Deep reinforcement learning based scheduling within production plan in semiconductor fabrication', *Expert Systems with Applications*, vol. 191, p. 116222.

Leijnen, S. and van Veen, F. (2019) 'The Neural Network Zoo', *IS4SI 2019 Summit*. Berkeley, CA, USA, 2–6 June 2019. Basel Switzerland, MDPI, p. 9.

Leo Kumar, S. P. (2017) 'State of The Art-Intense Review on Artificial Intelligence Systems Application in Process Planning and Manufacturing', *Engineering Applications of Artificial Intelligence*, vol. 65, pp. 294–329.

Li, C., Zheng, P., Yin, Y., Wang, B. and Wang, L. (2023) 'Deep reinforcement learning in smart manufacturing: A review and prospects', *CIRP Journal of Manufacturing Science and Technology*, vol. 40, pp. 75–101.

Li, F. and Du, Y. (2018) 'From AlphaGo to Power System AI: What Engineers Can Learn from Solving the Most Complex Board Game', *IEEE Power and Energy Magazine*, vol. 16, no. 2, pp. 76–84.

Li, L., Qiu, Q., Xiao, Z., Lin, Q., Gu, J. and Ming, Z. (2024) 'A Two-Stage Hybrid Multi-Objective Optimization Evolutionary Algorithm for Computing Offloading in Sustainable Edge Computing', *IEEE Transactions on Consumer Electronics*, vol. 70, no. 1, pp. 735–746.

Liao, X., Wang, Y., Xuan, Y. and Wu, D. (2020) 'AGV Path Planning Model based on Reinforcement Learning', *Proceedings 2020 Chinese Automation Congress (CAC2020)*. Shanghai, China, 6–8 November 2020. Piscataway, NJ, IEEE, pp. 6722–6726.

Lillicrap, T. P., Hunt, J., Pritzel, A., Heess, N., Erez, T., Tassa, Y., Silver, D. and Wierstra, D. (2015) *Continuous control with deep reinforcement learning* [Online]. Available at http://arxiv.org/abs/1509.02971 (Accessed 22 December 2024).

Lin, J., Shen, A., Wu, L. and Zhong, Y. (2024) 'Learning-based simulated annealing algorithm for unequal area facility layout problem', *Soft Computing*, vol. 28, no. 6, pp. 5667–5682.

Lin, L.-J. (1992) 'Self-improving reactive agents based on reinforcement learning, planning and teaching', *Machine Learning*, vol. 8, 3–4, pp. 293–321.

Liu, Q. and Meller, R. D. (2007) 'A sequence-pair representation and MIP-model-based heuristic for the facility layout problem with rectangular departments', *IIE Transactions*, vol. 39, no. 4, pp. 377–394.

Liu, Y., Wang, C., Zhao, C., Wu, H. and Wei, Y. (2024) 'A Soft Actor-Critic Deep Reinforcement-Learning-Based Robot Navigation Method Using LiDAR', *Remote Sensing*, vol. 16, no. 12, p. 2072.

Lobos, D. and Donath, D. (2010) 'The problem of space layout in architecture: A survey and reflections', *ARQUITETURA REVISTA*, vol. 6, no. 2, pp. 136–161.

Lust, T. and Teghem, J. (2010) 'The Multiobjective Traveling Salesman Problem: A Survey and a New Approach', *Advances in Multi-Objective Nature Inspired Computing*, vol. 272, pp. 119–141 [Online]. https://doi.org/10.1007/978-3-642-11218-8_6.

Madhusudanan Pillai, V., Hunagund, I. B. and Krishnan, K. K. (2011) 'Design of robust layout for Dynamic Plant Layout Problems', *Computers & Industrial Engineering*, vol. 61, no. 3, pp. 813–823.

Maganha, I. and Silva, C. (2017) 'A Theoretical Background for the Reconfigurable Layout Problem', *Procedia Manufacturing*, vol. 11, pp. 2025–2033.

Malus, A., Kozjek, D. and Vrabič, R. (2020) 'Real-time order dispatching for a fleet of autonomous mobile robots using multi-agent reinforcement learning', *CIRP Annals*, vol. 69, no. 1, pp. 397–400.

Marshall, J. and McKay, P. (2005) 'A Review of Design Science in Information System', *ACIS 2005 Proceedings*. Sydney, Australia, 29 Nov–2 Dec.

Martens, J. (2004) 'Two genetic algorithms to solve a layout problem in the fashion industry', *European Journal of Operational Research*, vol. 154, no. 1, p. 304.

McCulloch, W. S. and Pitts, W. (1943) 'A logical calculus of the ideas immanent in nervous activity', *The Bulletin of Mathematical Biophysics*, vol. 5, no. 4, pp. 115–133.

Meller, R. D. and Bozer, Y. A. (1996) 'A new simulated annealing algorithm for the facility layout problem', *International Journal of Production Research*, vol. 34, no. 6, pp. 1675–1692.

Meller, R. D. and Gau, K.-Y. (1996) 'The facility layout problem: Recent and emerging trends and perspectives', *Journal of Manufacturing Systems*, vol. 15, no. 5, pp. 351–366.

Miç, P. (2022) 'Comparative solution approach with quadratic and genetic programming to multi-objective healthcare facility location problem', *Sigma Journal of Engineering and Natural Sciences*, vol. 40, no. 1, pp. 63–78.

Miikkulainen, R. (1994) 'Integrated connectionist models: building AI systems on subsymbolic foundations', *Proceedings / Sixth International Conference on Tools with Artificial*

Intelligence, November 6–9, 1994, New Orleans, Louisiana. New Orleans, LA, USA, 6–9 Nov. 1994. Los Alamitos, Calif., IEEE Computer Society Press, pp. 231–232.

Mnih, V., Badia, A. P., Mirza, M., Graves, A., Lillicrap, T., Harley, T., Silver, D. and Kavukcuoglu, K. (2016) 'Asynchronous Methods for Deep Reinforcement Learning', *Proceedings of The 33rd International Conference on Machine Learning.* New York, New York, USA, June 19–24, 2016, pp. 1928–1937.

Mnih, V., Kavukcuoglu, K., Silver, D., Graves, A., Antonoglou, I., Wierstra, D. and Riedmiller, M. (2013) *Playing atari with deep reinforcement learning* [Online]. Available at https://people.engr.tamu.edu/guni/csce642/files/dqn.pdf (Accessed 21 December 2024).

Mnih, V., Kavukcuoglu, K., Silver, D., Rusu, A. A., Veness, J., Bellemare, M. G., Graves, A., Riedmiller, M., Fidjeland, A. K., Ostrovski, G., Petersen, S., Beattie, C., Sadik, A., Antonoglou, I., King, H., Kumaran, D., Wierstra, D., Legg, S. and Hassabis, D. (2015) 'Human-level control through deep reinforcement learning', *Nature*, vol. 518, no. 7540, pp. 529–533.

Moher, D., Liberati, A., Tetzlaff, J. and Altman, D. G. (2009) 'Preferred reporting items for systematic reviews and meta-analyses: the PRISMA statement', *Annals of Internal Medicine*, vol. 151, no. 4, 264–9, W64.

Momenikorbekandi, A. and Abbod, M. (2023) 'Intelligent Scheduling Based on Reinforcement Learning Approaches: Applying Advanced Q-Learning and State–Action–Reward–State–Action Reinforcement Learning Models for the Optimisation of Job Shop Scheduling Problems', *Electronics*, vol. 12, no. 23, p. 4752.

Monostori, L., Hornyák, J., Egresits, C. and Viharos, Z. J. (1998) 'Soft computing and hybrid AI approaches to intelligent manufacturing', in Del Pasqual Pobil, A., Mira, J. and Ali, M. (eds) *Tasks and Methods in Applied Artificial Intelligence: 11th International Conference on Industrial and Engineering Applications of Artificial Intelligence and Expert Systems IEA-98-AIE Benicàssim, Castellón, Spain, June 1–4, 1998 Proceedings, Volume II,* Berlin, Heidelberg, Springer Berlin Heidelberg, pp. 765–774.

Montreuil, B., Venkatadri, U. and Ratliff, H. D. (1993) 'Generating a Layout From a Design Skeleton', *IIE Transactions*, vol. 25, no. 1, pp. 3–15.

Moosmann, M., Kulig, M., Spenrath, F., Mönnig, M., Roggendorf, S., Petrovic, O., Bormann, R. and Huber, M. F. (2021) 'Separating Entangled Workpieces in Random Bin Picking using Deep Reinforcement Learning', *Procedia CIRP*, vol. 104, pp. 881–886.

Moritz, P., Nishihara, R., Wang, S., Tumanov, A., Liaw, R., Liang, E., Elibol, M., Yang, Z., Paul, W., Jordan, M. I. and Stoica, I. (2018) 'Ray: A Distributed Framework for Emerging AI Applications', *13th USENIX Symposium on Operating Systems Design and Implementation (OSDI 18).* Carlsbad, CA, USENIX Association, pp. 561–577.

Moslemipour, G., Lee, T. S. and Rilling, D. (2012) 'A review of intelligent approaches for designing dynamic and robust layouts in flexible manufacturing systems', *The International Journal of Advanced Manufacturing Technology*, vol. 60, pp. 11–27.

Muther, R. (1973) *Systematic Layout Planning*, 2nd edn, Boston, Cahners Books.

Newell, A. (1980) 'Physical Symbol Systems', *Cognitive Science*, vol. 4, no. 2, pp. 135–183.

Nicol, L. M. and Hollier, R. H. (1983) 'Plant layout in practice', *Material Flow*, vol. 1, no. 3, pp. 177–188.

Niebles, F., Escobar, I., Agudelo, L. and Jimenez, G. (2016) 'A Comparative Analysis of Genetic Algorithms and QAP Formulation for Facility Layout Problem: An Application in a Real Context', in Tan, Y., Shi, Y. and Li, L. (eds) *Advances in swarm intelligence:*

7th international conference, ICSI 2016, Bali, Indonesia, June 25–30, 2016 : proceedings, Cham, Heidelberg, Springer, pp. 59–75.

Niehaves, B. (2007) 'On Epistemological Pluralism in Design Science', *Scandinavian Journal of Information Systems*, vol. 19, no. 2, pp. 93–104 [Online]. Available at https://aisel.aisnet.org/cgi/viewcontent.cgi?article=1290&context=icis2007.

Niroomand, S., Hadi-Vencheh, A., Şahin, R. and Vizvári, B. (2015) 'Modified migrating birds optimization algorithm for closed loop layout with exact distances in flexible manufacturing systems', *Expert Systems with Applications*, vol. 42, no. 19, pp. 6586–6597.

Nugent, C. E., Vollmann, T. E. and Ruml, J. (1968) 'An Experimental Comparison of Techniques for the Assignment of Facilities to Locations', *Operations Research*, vol. 16, no. 1, pp. 150–173.

Odu, G. O. (2019) 'Weighting methods for multi-criteria decision making technique', *Journal of Applied Sciences and Environmental Management*, vol. 23, no. 8, p. 1449.

Ono, S., Hamada, Y., Yamamoto, S., Mizuno, K. and Nishihara, S. (2001) 'Rearrangement of floor layouts based on constraint satisfaction', *E-systems and e-man for cybernetics in cyberspace*. Tucson, AZ, USA, 7–10 October 2001. Piscataway, NJ, IEEE Service Center, pp. 2759–2764.

OpenAI, Akkaya, I., Andrychowicz, M., Chociej, M., Litwin, M., McGrew, B., Petron, A., Paino, A., Plappert, M., Powell, G., Ribas, R., Schneider, J., Tezak, N., Tworek, J., Welinder, P., Weng, L., Yuan, Q., Zaremba, W. and Zhang, L. (2019) *Solving Rubik's Cube with a Robot Hand* [Online]. Available at http://arxiv.org/pdf/1910.07113 (Accessed 12 May 2025).

Panzer, M. and Bender, B. (2022) 'Deep reinforcement learning in production systems: a systematic literature review', *International Journal of Production Research*, vol. 60, no. 13, pp. 4316–4341.

Park, J., Chun, J., Kim, S. H., Kim, Y. and Park, J. (2021) 'Learning to schedule job-shop problems: representation and policy learning using graph neural network and reinforcement learning', *International Journal of Production Research*, vol. 59, no. 11, pp. 3360–3377.

Patel, D., Hazan, H., Saunders, D. J., Siegelmann, H. T. and Kozma, R. (2019) 'Improved robustness of reinforcement learning policies upon conversion to spiking neuronal network platforms applied to Atari Breakout game', *Neural networks : the official journal of the International Neural Network Society*, vol. 120, pp. 108–115.

Paul, R. C., Asokan, P. and Prabhakar, V. I. (2006) 'A solution to the facility layout problem having passages and inner structure walls using particle swarm optimization', *The International Journal of Advanced Manufacturing Technology*, vol. 29, 7–8, pp. 766–771.

Pawellek, G. (2014) *Ganzheitliche Fabrikplanung: Grundlagen, Vorgehensweise, EDV-Unterstützung*, 2nd edn, Berlin, Heidelberg, Springer-Vieweg.

Pérez-Gosende, P., Mula, J. and Díaz-Madroñero, M. (2021) 'Facility layout planning. An extended literature review', *International Journal of Production Research*, vol. 59, no. 12, pp. 3777–3816.

Peron, M., Fragapane, G., Sgarbossa, F. and Kay, M. (2020) 'Digital Facility Layout Planning', *Sustainability*, vol. 12, no. 8, p. 3349.

Potocnik, P., Berlec, T., Sluga, A. and Govekar, E. (2014) 'Hybrid Self-Organization Based Facility Layout Planning', *STROJNISKI VESTNIK-JOURNAL OF MECHANICAL ENGINEERING*, vol. 60, no. 12, pp. 789–796.

Qi, L., Zeng, Q., Liu, S., Wang, J., Qin, S. and Guo, X. (2025) 'Twin Delayed Deep Deterministic Policy Gradient Algorithm for a Heterogeneous Multifactory Remanufacturing Optimization Problem', *IEEE Transactions on Computational Social Systems*, pp. 1–12.

Raffin, A., Hill, A. and Ernestus, M. (2019) *Stable Baselines3* [Online]. Available at https://github.com/DLR-RM/stable-baselines3 (Accessed 22 December 2024).

Raman, D., Nagalingam, S. V. and Gurd, B. W. (2009) 'A genetic algorithm and queuing theory based methodology for facilities layout problem', *International Journal of Production Research*, vol. 47, no. 20, pp. 5611–5635.

Rao, H. A. and Gu, P. (1995) 'A multi-constraint neural network for the pragmatic design of cellular manufacturing systems', *International Journal of Production Research*, vol. 33, no. 4, p. 1048.

Renzi, C., Leali, F., Cavazzuti, M. and Andrisano, A. O. (2014) 'A review on artificial intelligence applications to the optimal design of dedicated and reconfigurable manufacturing systems', *The International Journal of Advanced Manufacturing Technology*, vol. 72, pp. 403–418.

Ripon, K. S. N., Glette, K., Khan, K. N., Hovin, M. and Torresen, J. (2013) 'Adaptive variable neighborhood search for solving multi-objective facility layout problems with unequal area facilities', *Swarm and Evolutionary Computation*, vol. 8, pp. 1–12.

Rolf, B., Jackson, I., Müller, M., Lang, S., Reggelin, T. and Ivanov, D. (2023) 'A review on reinforcement learning algorithms and applications in supply chain management', *International Journal of Production Research*, vol. 61, no. 20, pp. 7151–7179.

Rosenblatt, M. J. (1986) 'The Dynamics of Plant Layout', *Management Science*, vol. 32, no. 1, pp. 76–86.

Rummery, G. and Niranjan, M. (1994) *On-line Q-learning using connectionist sytems*, (Technical Report CUED/F-INFENG-TR 166).

Rummukainen, H. and Nurminen, J. K. (2019) 'Practical Reinforcement Learning - Experiences in Lot Scheduling Application', *IFAC-PapersOnLine*, vol. 52, no. 13, pp. 1415–1420.

Russell, S. J. and Norvig, P. (2016) *Artificial intelligence: A modern approach*, 3rd edn, Boston, Pearson.

Saaty, T. L. (1977) 'A scaling method for priorities in hierarchical structures', *Journal of Mathematical Psychology*, vol. 15, no. 3, pp. 234–281.

Saaty, T. L. (1980) *The Analytic Hierarchy Process*, New York, McGraw-Hill.

Saaty, T. L. (1990) 'How to make a decision: The analytic hierarchy process', *European Journal of Operational Research*, vol. 48, no. 1, pp. 9–26.

Saaty, T. L. (2013) *Decision making with the analytic network process: Economic, political, social and technological applications with benefits, opportunities, costs and risks*, 2nd edn, New York, Springer.

Sahragard, M. and Bashiri, M. (2016) 'Layout design for large scale problems with a hybridized clustering based heuristic method', *IEEM 2016: International Conference on Industrial Engineering and Engineering Management : 4–7 December 2016, Bali, Indonesia*. Bali, Indonesia, 12/4/2016–12/7/2016. Piscataway, NJ, IEEE, pp. 1610–1614.

Salas-Morera, L., García-Hernández, L. and Carmona-Muñoz, C. (2021) 'A Multi-User Interactive Coral Reef Optimization Algorithm for Considering Expert Knowledge in the Unequal Area Facility Layout Problem', *Applied Sciences*, vol. 11, no. 15, p. 6676.

Sammut, C. and Webb, G. I. (2011) *Encyclopedia of machine learning*, New York, NY, Springer.

Samuel, A. L. (1959) 'Some Studies in Machine Learning Using the Game of Checkers', *IBM Journal of Research & Development*, vol. 3, no. 3, pp. 210–229.

Saßmannshausen, T. and Heinbach, B. (2025) 'Cyborgs in the Factory: How Blending Biology and Technology is Shaping the Future of Work', *Procedia CIRP*, Accepted.

Sauer, C. R. and Burggräf, P. (2024) 'Hybrid intelligence—systematic approach and framework to determine the level of Human-AI collaboration for production management use cases', *Production Engineering*.

Schäfer, L., Klenk, F., Maier, T., Zehner, M., Peukert, S., Linzbach, R., Treiber, T. and Lanza, G. (2024) 'A systematic approach for simulation-based dimensioning of production systems during the concept phase of factory planning', *Production Engineering*.

Schmigalla, H. (1970) *Methoden zur optimalen Maschinenanordnung*, VEB Verlag Technik.

Schneckenreither, M., Windmueller, S. and Haeussler, S. (2021) 'Smart Short Term Capacity Planning: A Reinforcement Learning Approach', in Dolgui, A., Bernard, A., Lemoine, D., Cieminski, G. von and Romero, D. (eds) *Advances in Production Management Systems: Artificial intelligence for sustainable and resilient production systems : IFIP WG 5.7 International Conference, APMS 2021, Nantes, France, September 5–9, 2021, proceedings, part I*, Cham, Springer International Publishing; Springer, pp. 258–266.

Schneidewind, J. and Galka, S. (2023) 'Interactive Reinforcement Learning-Based Factory Layout Planning', *13th Conference on Learning Factories 2023 (CLF)*. University of Reutlingen, Germany, 9–11 May, 2023. Amsterdam, Amsterdam Centre for European Studies.

Scholz, D. (2010) *Innerbetriebliche Standortplanung: Das Konzept der Slicing Trees bei der Optimierung von Layoutstrukturen* (Zugl.: Darmstadt, Techn.-Univ., Diss., 2010), Wiesbaden, Gabler.

Scholz, D., Jaehn, F. and Junker, A. (2010) 'Extensions to STaTS for practical applications of the facility layout problem', *European Journal of Operational Research*, vol. 204, no. 3, pp. 463–472.

Schrittwieser, J., Antonoglou, I., Hubert, T., Simonyan, K., Sifre, L., Schmitt, S., Guez, A., Lockhart, E., Hassabis, D., Graepel, T., Lillicrap, T. and Silver, D. (2020) 'Mastering Atari, Go, chess and shogi by planning with a learned model', *Nature*, vol. 588, no. 7839, pp. 604–609.

Schuh, G., Kampker, A. and Wesch-Potente, C. (2011) 'Condition based factory planning', *Production Engineering*, vol. 5, no. 1, pp. 89–94.

Schulman, J., Levine, S., Abbeel, P., Jordan, M. and Moritz, P. (2015) 'Trust Region Policy Optimization', *Proceedings of the 32nd International Conference on Machine Learning*. Lille, France, PMLR, pp. 1889–1897.

Schulman, J., Wolski, F., Dhariwal, P., Radford, A. and Klimov, O. (2017) *Proximal policy optimization algorithms* [Online]. Available at https://arxiv.org/pdf/1707.06347 (Accessed 21 December 2024).

Seehof, J. M. and Evans, W. O. (1967) 'Automated layout design program', *The Journal of Industrial Engineering*, vol. 18, pp. 690–695.

Serra, T. and O'Neil, R. J. (2020) 'MIPLIBing: Seamless Benchmarking of Mathematical Optimization Problems and Metadata Extensions', *SN Operations Research Forum*, vol. 1, no. 3.

Sharma, P. and Singhal, S. (2016) 'A review of objectives and solution approaches for facility layout problems', *International Journal of Industrial and Systems Engineering*, vol. 24, no. 4, p. 469.

Shi, D., Fan, W., Xiao, Y., Lin, T. and Xing, C. (2020) 'Intelligent scheduling of discrete automated production line via deep reinforcement learning', *International Journal of Production Research*, vol. 58, no. 11, pp. 3362–3380.

Shouman, M. A., Nawara, G. M., Reyad, A. H. and El-Darandaly, K. (2001) 'Facility layout problem (FLP) and intelligent techniques: a survey', *7th International Conference on Production Engineering, Design and Control.* Alexandria, Egypt, February 13–15.

Silver, D., Huang, A., Maddison, C. J., Guez, A., Sifre, L., van den Driessche, G., Schrittwieser, J., Antonoglou, I., Panneershelvam, V., Lanctot, M., Dieleman, S., Grewe, D., Nham, J., Kalchbrenner, N., Sutskever, I., Lillicrap, T., Leach, M., Kavukcuoglu, K., Graepel, T. and Hassabis, D. (2016) 'Mastering the game of Go with deep neural networks and tree search', *Nature*, vol. 529, no. 7587, pp. 484–489.

Silver, D., Hubert, T., Schrittwieser, J., Antonoglou, I., Lai, M., Guez, A., Lanctot, M., Sifre, L., Kumaran, D., Graepel, T., Lillicrap, T., Simonyan, K. and Hassabis, D. (2017) *Mastering Chess and Shogi by Self-Play with a General Reinforcement Learning Algorithm* [Online]. Available at http://arxiv.org/pdf/1712.01815.

Silver, D., Lever, G., Heess, N., Degris, T., Wierstra, D. and Riedmiller, M. (2014) 'Deterministic Policy Gradient Algorithms', *Proceedings of the 31st International Conference on Machine Learning.* Bejing, China, PMLR, pp. 387–395.

Silver, D., Schrittwieser, J., Simonyan, K., Antonoglou, I., Huang, A., Guez, A., Hubert, T., Baker, L., Lai, M., Bolton, A., Chen, Y., Lillicrap, T., Hui, F., Sifre, L., van den Driessche, G., Graepel, T. and Hassabis, D. (2017) 'Mastering the game of Go without human knowledge', *Nature*, vol. 550, no. 7676, pp. 354–359.

Simmons, D. M. (1969) 'One-Dimensional Space Allocation: An Ordering Algorithm', *Operations Research*, vol. 17, no. 5, pp. 812–826.

Singh, S. P. and Sharma, R. R. K. (2006) 'A review of different approaches to the facility layout problems', *The International Journal of Advanced Manufacturing Technology*, vol. 30, 5–6, pp. 425–433 [Online]. https://doi.org/10.1007/s00170-005-0087-9.

Spangher, L., Gokul, A., Palakapilly, J., Agwan, U., Khattar, M., Ma, W.-J. and Spanos, C. (2020) *Officelearn: an OpenAI Gym environment for reinforcement learning on occupant-level building's energy demand response* [Online]. Available at https://www.climatechange.ai/papers/neurips2020/56/paper.pdf (Accessed 22 December 2024).

Stricker, N., Kuhnle, A., Sturm, R. and Friess, S. (2018) 'Reinforcement learning for adaptive order dispatching in the semiconductor industry', *CIRP Annals*, vol. 67, no. 1, pp. 511–514.

Stuke, T., Rauschenbach, T. and Bartsch, T. (2024) 'Development of a Robotic Bin Picking Approach Based on Reinforcement Learning', in Niggemann, O., Beyerer, J., Krantz, M. and Kühnert, C. (eds) *Machine Learning for Cyber-Physical Systems: Selected papers from the International Conference ML4CPS 2023,* Cham, Springer Nature Switzerland; Imprint Springer, pp. 41–49.

Su, J., Huang, J., Adams, S., Chang, Q. and Beling, P. A. (2022) 'Deep multi-agent reinforcement learning for multi-level preventive maintenance in manufacturing systems', *Expert Systems with Applications*, vol. 192, p. 116323.

Sumanas, M., Petronis, A., Bucinskas, V., Dzedzickis, A., Virzonis, D. and Morkvenaite-Vilkonciene, I. (2022) 'Deep Q-Learning in Robotics: Improvement of Accuracy and Repeatability', *Sensors (Basel, Switzerland)*, vol. 22, no. 10.

Sun, X., Lai, L.-F., Chou, P., Chen, L.-R. and Wu, C.-C. (2018) 'On GPU Implementation of the Island Model Genetic Algorithm for Solving the Unequal Area Facility Layout Problem', *Applied Sciences*, vol. 8, no. 9, p. 1604.

Süße, M., Ahrens, A., Richter-Trummer, V. and Ihlenfeldt, S. (2023) 'Assisted Facility Layout Planning for Sustainable Automotive Assembly', *Future Automotive Production Conference 2022.* Wolfsburg, Germany, 17–18 May, 2022. Wiesbaden, Germany and Heidelberg, Springer Vieweg, pp. 173–188.

Sutton, R. S. (1990) 'Integrated Architectures for Learning, Planning, and Reacting Based on Approximating Dynamic Programming', in Porter, B. and Mooney, R. (eds) *Machine Learning Proceedings 1990: Proceedings of the Seventh International Conference on Machine Learning, University of Texas, Austin, Texas, June 21–23 1990,* s.l., Elsevier Reference Monographs, pp. 216–224.

Sutton, R. S. and Barto, A. (2020) *Reinforcement learning: An introduction*, Cambridge, Massachusetts, London, England, The MIT Press.

Sutton, R. S. and Barto, A. G. (2018) *Reinforcement learning: An introduction*, 2nd edn, Cambridge, Massachusetts, MIT Press.

Sutton, R. S., McAllester, D., Singh, S. and Mansour, Y. (1999) 'Policy Gradient Methods for Reinforcement Learning with Function Approximation', *Advances in Neural Information Processing Systems,* MIT Press.

Tam, C. M. and Tong, T. K. L. (2003) 'GA-ANN model for optimizing the locations of tower crane and supply points for high-rise public housing construction', *Construction Management & Economics*, vol. 21, no. 3, pp. 257–266.

Tam, K. Y. (1992) 'A simulated annealing algorithm for allocating space to manufacturing cells', *International Journal of Production Research*, vol. 30, no. 1, pp. 63–87.

Tarkesh, H., Atighehchian, A. and Nookabadi, A. S. (2009) 'Facility layout design using virtual multi-agent system', *Journal of Intelligent Manufacturing*, vol. 20, no. 4, p. 347.

Tateyama, T. and Kawata, S. (2004) 'Plant layout planning using ART neural networks', *SICE 2004 Annual Conference.* Sapporo, Japan, 4–6 August 2004, 1863–1868 vol. 2.

Tateyama, T., Kawata, S. and Ohta, H. (2003a) 'A conditional clustering algorithm using self-organizing map', *SICE 2003 Annual Conference.* Fukui, Japan, August 4–6, 2003. Piscataway, NJ, IEEE, 3259–3264 Vol.3.

Tateyama, T., Kawata, S. and Ohta, H. (2003b) 'Self-organizing map for Group Technology oriented plant layout planning', *IEICE Transactions on Fundamentals of Electronics Communications and Computer Sciences*, E86A, no. 11, pp. 2747–2754.

Tayal, A., Kose, U., Solanki, A., Nayyar, A. and Marmolejo Saucedo, J. A. (2020) 'Efficiency analysis for stochastic dynamic facility layout problem using meta-heuristic, data envelopment analysis and machine learning', *Computational Intelligence*, vol. 36, no. 1, pp. 172–202.

Thattai, K., Ravishankar, J. and Li, C. (2023) 'Consumer-Centric Home Energy Management System Using Trust Region Policy Optimization- Based Multi-Agent Deep Reinforcement Learning', *2023 IEEE Belgrade PowerTech: June 25–29, 2023, Belgrade, Serbia.* Belgrade, Serbia, 6/25/2023–6/29/2023. Piscataway, NJ, IEEE, pp. 1–6.

Tompkins, J., White, J. A. and Bozer, Y. A. (2010) *Facilities planning*, 4th edn, Hoboken, NJ, Wiley.

Tong, X. (1991) *SECOT: a sequential construction technique for facility design* [Online], Pittsburgh, University of Pittsburgh. Available at https://search.proquest.com/openview/fce93f2ee4d281294a551da4d9dbbfe2/1?pq-origsite=gscholar&cbl=18750&diss=y (Accessed 22 December 2024).

Tseng, C.-Y., Li, J., Lin, L.-H., Wang, K., White III, C. C. and Wang, B. (2025) 'Deep reinforcement learning approach for dynamic capacity planning in decentralised regenerative medicine supply chains', *International Journal of Production Research*, vol. 63, no. 2, pp. 555–570.

Tsuchiya, K., Bharitkar, S. and Takefuji, Y. (1996) 'A neural network approach to facility layout problems', *European Journal of Operational Research*, vol. 89, no. 3, pp. 556–563.

Turing, A. (1948) *Intelligent machinery*, Berlin, Springer-Verlag.

Ueda, K., Fujii, N., Hatono, I. and Kobayashi, M. (2002) 'Facility Layout Planning Using Self-Organization Method', *CIRP Annals*, vol. 51, no. 1, pp. 399–402.

Ulutas, B. and Islier, A. A. (2015) 'Dynamic facility layout problem in footwear industry', *Journal of Manufacturing Systems*, vol. 36, pp. 55–61.

Ulutas, B. H. and Kulturel-Konak, S. (2012) 'An artificial immune system based algorithm to solve unequal area facility layout problem', *Expert Systems with Applications*, vol. 39, no. 5, pp. 5384–5395.

Unger, H. and Börner, F. (2021) 'Reinforcement Learning for Layout Planning—Modelling the Layout Problem as MDP', in Dolgui, A., Bernard, A., Lemoine, D., Cieminski, G. von and Romero, D. (eds) *Advances in Production Management Systems: Artificial intelligence for sustainable and resilient production systems : IFIP WG 5.7 International Conference, APMS 2021, Nantes, France, September 5–9, 2021, proceedings, part I*, Cham, Springer International Publishing; Springer, pp. 471–479.

Unger, H., Börner, F. and Fischer, D. (2024) 'Reinforcement Learning for Layout Planning—Automated Pathway Generation for Arbitrary Factory Layouts', in Silva, F. J. G. (ed) *Flexible Automation and Intelligent Manufacturing: Proceedings of FAIM 2023, June 18–22, 2023, Porto, Portugal, Volume 2: Industrial Management*, Cham, Springer, pp. 1031–1039.

Vakharia, A. J. and Wemmerlöv, U. (1995) 'A comparative investigation of hierarchical clustering techniques and dissimilarity measures applied to the cell formation problem', *Journal of Operations Management*, vol. 13, no. 2, pp. 117–138.

Vázquez-Canteli, J. R., Kämpf, J., Henze, G. and Nagy, Z. (2019) 'CityLearn v1.0', *BuildSys '19: The 6th ACM International Conference on Systems for Energy-Efficient Buildings, Cities, and Transportation*. New York, NY, USA, November 13–14, 2019. New York, NY, United States, Association for Computing Machinery, pp. 356–357.

VDI (2011) *VDI 5200 Blatt 1: Fabrikplanung–Planungsvorgehen*, Berlin, Beuth Verlag.

Vinyals, O., Babuschkin, I., Chung, J., Mathieu, M., Jaderberg, M. and Czarnecki, W. M. (2019) *Alphastar: Mastering the real-time strategy game starcraft ii* [Online]. Available at https://deepmind.com/blog/article/alphastar-mastering-real-time-strategy-game-starcraft-ii (Accessed 21 December 2024).

vom Brocke, J. (2020) *Design Science Research. Cases*, Cham, Springer International Publishing AG.

Waschneck, B., Reichstaller, A., Belzner, L., Altenmüller, T., Bauernhansl, T., Knapp, A. and Kyek, A. (2018) 'Optimization of global production scheduling with deep reinforcement learning', *Procedia CIRP*, vol. 72, pp. 1264–1269.

Watkins, C. J. C. H. and Dayan, P. (1992) 'Q-learning', *Machine Learning*, vol. 8, 3–4, pp. 279–292.

Weber, T., Racanière, S., Reichert, D. P., Buesing, L., Guez, A., Rezende, D. J., Badia, A. P., Vinyals, O., Heess, N., Li, Y., Pascanu, R., Battaglia, P., Hassabis, D., Silver, D. and Wierstra, D. (2017) *Imagination-Augmented Agents for Deep Reinforcement Learning* [Online]. Available at http://arxiv.org/pdf/1707.06203.

Weichert, D., Link, P., Stoll, A., Rüping, S., Ihlenfeldt, S. and Wrobel, S. (2019) 'A review of machine learning for the optimization of production processes', *The International Journal of Advanced Manufacturing Technology*, vol. 104, 5–8, pp. 1889–1902.

Weigold, M., Ranzau, H., Schaumann, S., Kohne, T., Panten, N. and Abele, E. (2021) 'Method for the application of deep reinforcement learning for optimised control of industrial energy supply systems by the example of a central cooling system', *CIRP Annals*, vol. 70, no. 1, pp. 17–20.

Weißer, T., Saßmannshausen, T., Ohrndorf, D., Burggräf, P. and Wagner, J. (2020) 'A clustering approach for topic filtering within systematic literature reviews', *MethodsX*, vol. 7, no. 100831, pp. 1–10.

Weitzel, T. and Glock, C. H. (2018) 'Energy management for stationary electric energy storage systems: A systematic literature review', *European Journal of Operational Research*, vol. 264, no. 2, pp. 582–606.

Wiendahl, H.-H., Reichardt, J. and Nyhuis, P. (2024) *Handbuch Fabrikplanung: Konzept, Gestaltung und Umsetzung wandlungsfähiger Produktionsstätten*, 3rd edn, München, Hanser Verlag.

Williams, R. J. (1992) 'Simple statistical gradient-following algorithms for connectionist reinforcement learning', *Machine Learning*, vol. 8, 3–4, pp. 229–256.

Wuest, T., Weimer, D., Irgens, C. and Thoben, K.-D. (2016) 'Machine learning in manufacturing: advantages, challenges, and applications', *Production & Manufacturing Research*, vol. 4, no. 1, pp. 23–45.

Xiao, Y., Xie, Y., Kulturel-Konak, S. and Konak, A. (2017) 'A problem evolution algorithm with linear programming for the dynamic facility layout problem—A general layout formulation', *Computers & Operations Research*, vol. 88, pp. 187–207.

Yang, T. and Peters, B. A. (1998) 'Flexible machine layout design for dynamic and uncertain production environments', *European Journal of Operational Research*, vol. 108, no. 1, pp. 49–64.

Yang, T., Su, C.-T. and Hsu, Y.-R. (2000) 'Systematic layout planning: a study on semiconductor wafer fabrication facilities', *International Journal of Operations & Production Management*, vol. 20, no. 11, pp. 1359–1371.

Yeh, I.-C. (2006) 'Architectural layout optimization using annealed neural network', *Automation in Construction*, vol. 15, no. 4, pp. 531–539.

Yu, J. and Guo, P. (2020) 'Run-to-Run Control of Chemical Mechanical Polishing Process Based on Deep Reinforcement Learning', *IEEE Transactions on Semiconductor Manufacturing*, vol. 33, no. 3, pp. 454–465.

Zahedi-Seresht, M., Sadeghi Bigham, B., Khosravi, S. and Nikpour, H. (2024) 'Oil Production Optimization Using Q-Learning Approach', *Processes*, vol. 12, no. 1, p. 110.

Zamora, I., Gonzalez Lopez, N., Vilches, V. M. and Cordero, A. H. (2016) *Extending the OpenAI Gym for robotics: a toolkit Extending the OpenAI Gym for robotics: a toolkit for reinforcement learning using ROS and Gazebo* [Online]. Available at https://arxiv.org/pdf/1608.05742.pdf (Accessed 22 December 2024).

Zeng, Q., Chen, Y., Yuan, R. and He, M. (2023) *Optimization method of unequal area dynamic facility layout under fixed-point movement strategy.*

Zhang, G. W., Zhang, S. C. and Xu, Y. S. (2002) 'Research on flexible transfer line schematic design using hierarchical process planning', *Journal of Materials Processing Technology*, vol. 129, 1–3, pp. 629–633.

Zhang, Y. and Ye, W. (2019) 'Deep learning-based inverse method for layout design', *Structural and Multidisciplinary Optimization*, vol. 60, no. 2, pp. 527–536.

Zhang, Z., Chen, J. and Zhao, W. (2024) 'Multi-AGV route planning in automated warehouse system based on shortest-time Q-learning algorithm', *Asian Journal of Control*, vol. 26, no. 2, pp. 683–702.

Zhao, X., Zhao, H., Chen, P. and Ding, H. (2020) 'Model accelerated reinforcement learning for high precision robotic assembly', *International Journal of Intelligent Robotics and Applications*, vol. 4, no. 2, pp. 202–216.

Zhao, Y. and Duan, D. (2024) 'Workshop Facility Layout Optimization Based on Deep Reinforcement Learning', *Processes*, vol. 12, no. 1, p. 201.

Zhu, D., Yang, B., Liu, Y., Wang, Z., Ma, K. and Guan, X. (2022) 'Energy management based on multi-agent deep reinforcement learning for a multi-energy industrial park', *Applied Energy*, vol. 311, p. 118636.

Zuniga, E. R., Moris, M. U., Syberfeldt, A., Fathi, M. and Rubio-Romero, J. C. (2020) 'A Simulation-Based Optimization Methodology for Facility Layout Design in Manufacturing', *IEEE Access*, vol. 8, pp. 163818–163828.

The manufacturer's authorised representative in the EU is Springer Nature Customer Service Centre GmbH, Europaplatz 3, 69115 Heidelberg, Germany. If you have any concerns regarding our products, please contact ProductSafety@springernature.com

Printed and bound by CPI Group (UK) Ltd, Croydon, CR0 4YY

07/07/2026

02160924-0001